Cosmology and Architecture in Premodern Islam

SUNY series in Islam
Seyyed Hossein Nasr, editor

Cosmology and Architecture in Premodern Islam

An Architectural Reading of Mystical Ideas

Samer Akkach

State University of New York Press

Published by
State University of New York Press, Albany

Printed in the United States of America

For information, contact State University of New York Press, Albany, NY
www.sunypress.edu

Production by Marilyn P. Semerad
Marketing by Michael Campochiaro

Library of Congress Cataloging-in-Publication Data

Akkach, Samer.
Cosmology and architecture in premodern Islam : an architectural reading of mystical ideas / Samer Akkach.
p. cm. — (SUNY series in Islam)
Includes bibliographical references and index.
IBSN 0-7914-6411-3 (hardcover : alk. paper)—ISBN 0-7914-6412-1 (pbk. : alk. paper)
1. Islamic art and symbolism. 2. Symbolism in architecture. 3. Architecture, Islamic. 4. Islamic cosmology. 5. Sufism. I. Title. II. Series.
BP182.5.A34 2005
726'.2'01—dc22

2004014267

10 9 8 7 6 5 4 3 2 1

Contents

Illustrations

Photos

Figures

Note to the Reader

Unless otherwise stated, all translations from Arabic sources are mine. Where translations of the same texts by others have been consulted or used, appropriate reference is given in the notes.

Unless otherwise stated, all translations from the Quran are adapted from those of M. Pickthall, A. J. Araburry, and N. J. Dawood and A. Y. Ali. The adaptations vary in extent according to the demands of the interpretive context and consistency.

In transliterations I have followed the system of *IJMES*. Diacritical marks are used consistently on italicized technical terms and book titles and where appropriate on personal and place names. Except where *h* indicates *hijrī* date, all dates are of the Common Era (CE).

Due to space limitation, I have identified sources in the notes by their date of publication. Full details are given in the selected bibliography. For ease of recognition, I have identified premodern Arabic sources by short or abbreviated titles.

Preface

When I was a little boy I used to love the snowfall in Damascus. Playing with friends in the street was fun, of course, but the real joy was in the gazing trick I had discovered and thought no one knew. Raised on a couch placed under the kitchen's window that opens onto a large light well, I used to stand up motionless gazing at the snow flakes silently and gracefully falling down. In a magical moment, as I concentrated hard on the falling motion, the situation switched: the snow flakes suddenly became still, and I began to rise. I knew it was a mere illusion, for the moment I blinked I was immobilized and had to start again, yet the sensation of rising was still real and exhilarating. For hours I used to play this gazing trick. The heavier the snow fell the faster I rose, and the harder I concentrated the smoother my ride. Perplexed by my untypical stillness, my parents used to be curious about what I was plotting, and I used to enjoy seeing the signs of disbelief on their faces when I innocently replied: "I am watching the snow!" Beside the snow there were only tall, smudgy concrete walls, so I could not be lying, but they could not figure out what I was doing. The delightful trick remained one of my well-kept secrets. In some long stretches of concentration, as the rising sensation sank deep into my body, the ascension felt monotonously endless, as if I were silently floating in an infinite space. I often wished it were real. I was curious to know what lies beyond the sky, the beautiful blue border of my world. I had a suppressed desire to take a look at God's fearful yet, surely, wondrous land. I wanted to see where God lives! I knew my gazing trick would never take me there, but I was sure that where God lives was a different world. As I grew up, I learned, with some disappointment, that our world extends infinitely beyond the blue sky and that we do not really know its bounds, let alone what lies beyond. From the Big Bang on, I came to learn, the universe has never ceased expanding, and since all began some billions of years ago, the world's immensity is beyond the humble human imagination to grasp. So even if my childish trick was imbued with some magical powers, I thought, my long hard gaze would still lead me nowhere. Many years later, as I became interested in premodern cosmology, I was pleasantly surprised to find out that my childish reveries were not all baseless fantasies. The world might have been much larger than what my little mind could comprehend, but for premodern cosmologists it was still a bounded world. Not only that, but beyond the known bounds another unfolds, one everlasting with wondrous landscape, luminous structures, and spiritual beings. Fascinating though they may be, all seemed as nice fairy tales, good for bedtime stories but wildly remote to the scientific concerns of the modern world. I studied architecture, an earthly bound, technical discipline, or so it seemed, which from the Damascene perspective appeared to have hardly anything more to it than commodity, firmness, and delight. A strange set of circumstances then took me to Australia, a mysterious land the remoteness of which my friends used to aptly describe, using a colloquial Damascene expression, as "just before God." Indeed, for a young man who had never traveled outside Damascus, Australia was as remote and imaginary as heavenly Jerusalem and the Land of Reality. Yet, it was in Sydney where I met two inspirational teachers, Peter Kollar and Adrian Snodgrass, whose interest in medieval metaphysics, cosmology, and symbolism immediately connected all the dots, bringing my disparate fields of interest into focus. Since then I have been set on the path that eventually led to the writing of this book. Through architecture I found myself able to explore the complexities of the bounded universe—its design, order, and meanings—and to examine how cosmological thinking mediates human acts of making and space ordering. Yet the journey into premodern Islamic cosmology, mysticism, and architecture has been long, challenging, and solitary. The text was written at different stages, and in hindsight if I am to embark on the same project afresh I would probably tackle it differently. Along the journey many people have influenced, supported, and facilitated my work. I am deeply indebted to them. Peter Kollar sparked my interest in the topic. Adrian Snodgrass supervised my doctoral research and had the profoundest influence on my thinking. He taught me how to make sense of premodern concepts, how to appreciate their significance, and, most important, how to remain relevant to the concerns of the modern world. Ahmad Shboul helped me develop my translation and interpretive skills and expanded my knowledge of the Islamic tradition. His mentorship and friendship over the years have been invaluable. Seyyed Hossein Nasr's generous advice and support have been instrumental in making me believe in the merits of my work, while Yasser Tabbaa's comments on an early draft have led to many improvements. Two anonymous readers provided valuable critiques and praise that led to many refinements. The AKPIA's visiting fellowship program at MIT allowed me to access the valuable resources at MIT and Harvard, and the warm and generous support of Nasser Rabbat has made my visit all the more memorable. Gülru Necipoğlu directed my attention to relevant Ottoman texts and afforded me several enjoyable discussions. Special thanks are due to many people in Maktabat al-Asad in Damascus, especially those in the manuscript and photocopy sections, for their continued help and support over the years; to AnnaLee Pauls from Princeton University Library for her assistance in acquiring copies of Arabic manuscripts; and to Sin Jee Li for preparing the graphic materials. Finally, my deepest gratitude goes to my wife and two sons whose love, care, and support have sustained my interest in the project. They kept faith in me despite my depriving them of precious family times and despite their tacit doubts about whether this book will ever be finished.

Introduction

In his famous *Iḥyā' ʿUlūm al-Dīn* (Reviving the Sciences of Religion), al-Ghazālī (d. 1111), the celebrated Muslim theologian and mystic, cites an intriguing analogy. He says: "As an architect draws (*yuṣawwir*) the details of a house in whiteness and then brings it out into existence according to the drawn exemplar (*nuskha*), so likewise the creator (*fāṭir*) of heaven and earth wrote the master copy of the world from beginning to end in the Preserved Tablet (*al-lawḥ al-maḥfūẓ*) and then brought it out into existence according to the written exemplar."[1] In many ways, this book is a commentary on, and an exposition of, this statement, an attempt to explore the philosophical and theological contexts that give sense to such an analogy in premodern Islam. In broad terms, the book is concerned with the question of art and religion, creativity and spirituality, with how religious thought and ideas can provide a context for understanding the meanings of human design and acts of making. In specific terms, it is concerned with the cosmological and cosmogonic ideas found in the writings of certain influential Muslim mystics and with their relevance to architecture and spatial organization. The intent, as the title suggests, is to construct a new interpretive context that enables an architectural reading of mystical ideas. By this I mean using a tendency in spatial ordering traceable in buildings, settlements, and landscapes as a tool to frame mystical literature and to organize cosmological ideas into a coherent whole. Within this frame, a complex conjunction of metaphysics, cosmology, and mysticism is constructed and brought to bear on tectonic expressions.

Central to this interpretive approach is the notion of *spatial sensibility*, understood as a particular awareness of space and a predisposition toward spatial organization shaped by a complex, multilayered worldview. As an untheorized and unaestheticized predilection or bias, spatial sensibility mediates between layered cosmological, geographical, and bodily conceptions and deliberate spatial ordering. Throughout the Islamic world a tendency to order spaces according to a cruciform layout is traceable in a sufficient number of examples to suggest ubiquity and consistency across temporal, geographical, and cultural distances. True, the fleshing out of this cryptic order reveals very rich stylistic variations across time and geography, but the order itself remains noticeably consistent. What lies beneath this consistent sense of spatial ordering is the core question this study is attempting to address. Deep beneath local sociopolitical and cultural conditions, I argue, lies a predisposition anchored in a wider religio-cosmological conception that manifests itself in a spatial sensibility traceable through modes of spatial ordering.

Hinged on the ways in which the cosmos, the geography, and the human body—their structure and interconnectedness—are conceived and described, spatial sensibility is naturally determined by prevailing scientific understanding and technological abilities. Today, our modern spatial sensibility is shaped by the remarkable scientific and technological achievements that have taken place over the past five hundred years. The discovery of the heliocentric system, the invention of perspective, the camera, and the computer, the development of modern physics (quantum and astrophysics) along with space and telecommunication technologies, and, more recently, the emergence of cyberspace, have all led to significant transformations in our spatial sensibility. In the last two decades, many illuminating studies have provided valuable insights into these transformations, meticulously mapping the critical shifts in human perceptions of reality.[2]

In contrast to the current understanding of a boundless and infinitely expanding universe, premodern Muslims thought of and described the cosmos as being finite, bounded, and with astronomically definable limits. The entirety of the cosmos was graspable by means of geometry, numbers, and the alphabet. It was conceived in the form of concentric circles, at the center of which humans dwelled and at the outer limit stood the all-encompassing divine Throne. Space and time, as we know them, terminated at the divine Throne, which formed the threshold into the divine realms of being. There and beyond, different modalities of space and time prevailed. The Throne, the outer limit of the universe, was also visualized to be "quadrangular" in form but with a sense of spatiality that was distinct from our own. Marking a transitional zone, the Throne was seen to partake in both the physical and metaphysical worlds. Within this geometrically defined and ordered cosmos things were interrelated; they occupied definite positions within an intricate hierarchy. Nothing stood in isolation or ambiguity; everything was carefully positioned. Premodern Islamic sources provide very detailed descriptions of the cosmos, and the Sufis, among other Muslim thinkers, present numerous geometrical diagrams that illustrate the fundamental order of being and the basic design of the world, translating into visual idioms the "textual" contents of al-Ghazālī's exemplar. This book examines one set of such diagrams. Most premodern cosmograms were circular and symmetrically ordered. The world was clearly round and its *blueprint* revealed symmetry and hierarchy. This kind of holistic conception is not possible today. The cosmos has become too complex to be represented in simple geometrical diagrams. But most important, physical space has now occupied the entire universe, the whole of reality, leaving no room for the cosmic entities that once defined the heavenly landscape, such as the divine Throne, the celestial Gardens, the Land of Reality, and the Cities of Light.

The finitude, graspability, and transparency of the cosmic structure have made it possible to think of the world in allegorical terms and to compare its genesis and structure to other natural and human-made objects, such as the human body, earth's geography, and the form of a building or a city. Premodern Islamic texts are replete with such examples, revealing a mode of reasoning that integrates the spatiality of the cosmos, of the geography, of the human body, and of the built form in an effective way. It is this sense of integrated spatiality that brings cosmology, geography, the human body, and architecture together, allowing them to be seen in terms of one another without the need of theoretical mediation. This is what this study aims to show. Today, this integrated spatiality has been fragmented, with each element assuming a world of its own. The core metaphysical order that used to unite the whole has disintegrated. As a result, cosmological ideas have lost their immediacy in people's lives and spatial practices, while being complexified with layers of impenetrable mathematics. By dwelling on premodern Islamic cosmology, this study highlights, if indirectly, the dramatic reshaping of our spatial sensibility by modern science and technology.

This reshaping naturally distances us from the operative context of premodern cosmology. In dealing with premodern texts and monuments I, therefore, make no claims of attempting to reconstruct the past *objectively* or to relate what *really* happened. My intent is to present a probable reading—influenced as it may be by modern preoccupations—that correlates texts and objects and highlights certain conjunctions of ideas and forms within a mystical frame of reference. The analyses are of interest to those concerned with the intellectual and artistic traditions of Islam, shedding fresh light on a complex area still requiring further development.

Scope and Challenges

The defining parameters of the study—premodern Islamic cosmology, architecture, and mysticism—present many insurmountable challenges both conceptually and methodologically. While grappling with these challenges, I do not pretend to have succeeded in finding satisfactory answers. With the broad interpretive scope adopted, I have had to live with some of their unsettling implications. First, Islamic cosmology is a vast and complex field that incorporates rich scientific, religious, and folkloric dimensions. Reducing such complexity to a single scheme constructed from a small sample of texts and presented out of a specific historical context is problematic. The same applies to Islamic mysticism, which in addition presents the problem of its tenuous relationship with mainstream Islam. Yet, this reductive tendency is inevitable for an interdisciplinary study such as this, which attempts to trace in a small volume significant crossings in three large fields of studies. Focusing on ideas and modes of thinking, I chose to work within a loosely defined historical context, extending from the tenth century to the eighteenth century. While this may concur with the common temporal definition of the "premodern" in the Islamic world, I am aware of the distortions involved in isolating "premodern" Muslims from the concurrent processes of modernity taking place in Europe since the sixteenth century, as well as in ignoring the continuity of premodern conceptions and practices after the eighteenth century. Within this broad historical context, my reading is guided by the writings of a core figure, the celebrated Sufi master Muḥyī al-Dīn Ibn al-ʿArabī (d. 1240), who is recognized for his imposing presence and far-reaching influence throughout the Islamic world. Ibn ʿArabī's ideas and teachings did not emerge in vacuum, however, but were preceded by a rich tradition. Therefore, his ideas are traced in the writings of earlier and later Sufis in order to show the continuity and consistency of certain cosmological conceptions in medieval Islam.[3]

The relationship between architecture and Sufism poses another challenge, one that is centered on the question of how widely were sophisticated Sufi ideas shared among lay people in general and artists and craftsmen in particular. Relevant though it may seem, the sources and line of research required to address this question are different to those pursued in this study. In my exploratory exercise, I am bringing together two sets of historical objects: built forms and texts. In the first I identify a recurring spatial order, and in the second I trace a set of cosmological ideas. Although both are the products of the same milieu of premodern Islam, imprecise as this may be, the relationship between them does not hinge primarily on the historical evidence that links them to one another but on the logic of the relationship and the agency of the spatial sensibility. The agency of spatial sensibility is far more complex than tracing a link between texts and objects, of course, although the latter is by no means a simple task.

In most premodern Islamic sources that refer, in one way or another, to architecture, one encounters a degree of *ambivalence* in the way architecture is engaged. They reveal simultaneous attraction to, and disinterest in, architecture while being preoccupied with issues of different significance, such as the events of a journey, the biography of a governor, some myths and anecdotes, or certain religious, mystical, or scientific matters. For example, Ikhwān al-Ṣafāʾ, the tenth-century group of mystics and philosophers, make numerous references to the creative and productive process of building and making; the tenth- and eleventh-century physicist and optician Ibn al-Haytham deals with vision, visuality, sense perception, and the aesthetic qualities of objects; and the twelfth-century traveler Ibn Jubayr gives extensive descriptions of the buildings and places he visits. Yet neither sustains any focused discussion on architecture itself. The Ikhwān even present a detailed analysis of the noble proportions of the human body, strikingly similar to Vitruvius' discussion of proportions in the temple and the human body, but as a chapter in their treatise on music. Considering that no medieval Islamic sources on architecture, similar to those by Vitruvius or Alberti, have reached us as well as the general lack of information about architects, one cannot speak with certainty about theory or intentionality in architectural design, and much of the hermeneutical interpretations of Islamic architecture remain in the domain of conjecture.

As regards my interpretive strategy, I am approaching the mystical writings on cosmology with architectural preoccupations, with *order* and *correspondence* forming the main lines of interdisciplinary crossing. Using the theme of *symbolism* I aim to project a certain trajectory of mystical ideas concerning the notion of *ordering* within both the divine and human contexts. Symbolism, as understood in this study, provides the main conceptual tools that enable one's mind to journey between the divine and human domains and to maneuver through the multiple states of being they involve. This trajectory inscribes its path by reference to *patterns of correspondence* that link these multiple states of being, revealing what might be described as "structural resonance" between the divine and human levels of existence. The basis of this *resonance,* as will be reconstructed from original sources, is a universal order conceived by the Sufis, as well as other religious authorities, to be the divine model governing all modes of manifestation and creation. It is the original structure according to which the universe was designed and laid out. Architectural ordering seen in this light, *re-presents*, as al-Ghazālī's analogy suggests, the divine order, adding a human-made layer to this hierarchy.

The principles of the universal order are traced through the religio-philosophical reasoning of how Being emerged from non-Being, and how original Unity gave birth to an inexhaustible multiplicity. Here I explore specifically the generative "move" from *unity* to *triplicity* and *quadrature*, seen as a central cosmogonic paradigm of simultaneous proliferation and synthesis. The move is explored in a variety of contexts and manifestations. The first trace of this move unfolds the metaphysial order, which is then traced in the cosmic order, which is in turn traced in the architectural order. Spatially, the move refers to the deployment of space from a central point along the three axes of what the French philosopher and metaphysician René Guénon describes as the "three-dimensional cross."[4] This study shows how this conception formed the cornerstone of spatial sensibility in premodern Islam. It also shows how the manifold manifestations and interrelatedness of this primary spatial order unfold a complex web of meanings and intricate patterns of correspondence that at once govern the world and materialize the order inscribed in the divine exemplar.

The Question of Difference

Dealing with "Islamic architecture," a category forged by nineteenth-century European orientalists, demands some reflections on how one defines and deals with the "Islamic." Whether implicit or clearly stated, the question of difference has long been at the heart of most attempts that aim to define the Islamic in historical, formal, spiritual, or cultural terms. Problematic though it may be, the question of difference has been vigorously pursued in almost all aspects of life throughout the Islamic world. Globalization along with the recent confrontations between Islam and the West has intensified the postcolonial search for identity, resulting in desperate attempts to define and repossess this difference. While nineteenth- and early twentieth-century orientalists were preoccupied with the unifying characteristics of an assumed homogeneity of Islam, recent scholarship has emphasized the richness and diversity of each and every aspect of a heterogenous Islam, including art and architecture. Inspired by the postmodern critique, many studies in the last two decades have contributed to deconstructing the Islamic by showing that it can no longer designate a monolithic, unchanging phenomenon at any level, not even the religious. With the heterogeneity of Islam becoming the central premise of academic inquiry in the field, speaking of universalities and consistencies runs the risk of essentialism: today's ultimate and indefensible "sin"! Yet the deconstruction of the Islamic has only complexified the question without offering viable alternatives. The question of what is Islamic architecture remains unanswerable.

While dealing with the Islamic as revealed through architectural expressions and modes of thinking, the way in which I have approached difference is somewhat different. Instead of seeking the salient characteristics of the Islamic, my main concern here is how can the architecture of premodern Islam help us understand our present conditions? How can it enable us to penetrate into worlds of meanings that seem completely closed to contemporary architects? These questions lead to a preoccupation with *what* and *how we can learn* from the *difference* that separates us, as modern subjects, from medieval objects. Accordingly, while dealing with the past, the accent is placed on the implications of this dealing for the here and now of our engagements. The choice to focus on the Sufi teachings and ideas is made not because they provide direct answers to our current problems or reveal modes of living capable of restoring or repossessing a lost identity but because they enable us to conceive of a significant possibility of being, one wherein architecture can be seen to interconnect intrinsically with all aspects of being, at once enriching and being enriched by peoples' modes of living and thinking, yet without constituting an end in itself or being constrained by rigid categories. Thus, this study is concerned, if indirectly, with rethinking the relationship between peoples and their built environments, and in this context, architectural symbolism is used to introduce a broader ontological context for the exploration of this relationship, one that extends far beyond the limited concerns of current architectural education and practice.

Thus by foregrounding a primary spatial order against rich scenes of tectonic embodiments, it is not my intention to argue for or against the Islamic identity or "ownership" of this order. In fact this primary spatial order is common in many premodern traditions.[5] My aim is to explore a probable trajectory of meanings, not to search for evidence of identity, a trajectory that enables a fresh reading of premodern Islamic architecture from nonarchitectural sources. Veiled by layers of mystical and metaphysical interpretations, this reading is not preoccupied with formal, stylistic, or aesthetic qualities but rather with the intricacies of the conditions of being. In this context architecture becomes the lens through which humans see and understand the universalities that complement the cultural workings of their earthly existence.

In this study I hope to offer an alternative perspective to view the architecture of premodern Islam, one that complements existing approaches in a meaningful way. By shifting the focus away from style and history to ontology and cosmology, architecture can become a useful tool to access new literature, engage different sources, and organize knowledge about profound topics, rather than being the prime target of explanation. By this shift I aim to use architecture to make the reader aware of certain patterns of thought within the premodern Islamic tradition, instead of the normal scenarios where conceptual patterns are constructed to explain the nature and particularity of architecture. This has two advantages: first, shifting the focus away from architecture itself liberates architectural forms from the burden of historicity and causal interpretation, that is, finding causes (including meanings) to explain formal qualities; second, it enables one to access a wider spectrum of literary material, breaks disciplinary boundaries, and unfolds new interpretations. This approach tends to emphasize the cogency and significance of the constructed narratives, whereby architecture becomes a suitable tool to understand the working of a premodern spatial sensibility and its coherent cosmology.

Abbreviations

ATR	*Architectural Theory Review*
Bayān	"Bayān Miqdār al-Sana al-Sarmadiyya . . . ," al-Qāshānī (MS. 5258 Ẓah.)
BJMES	*British Journal of Middle Eastern Studies*
BSOAS	*Bulletin of the School of Oriental and African Studies*
EI2	*Encyclopaedia of Islam*, 2nd ed.
Fus.	*Fuṣūṣ al-Ḥikam*, 1: text by Ibn ʿArabī; 2: commentary by A. A. ʿAfīfī.
Fut.	*Al-Futūḥāt al-Makkiyya*, Ibn ʿArabī, Dār Ṣādir ed.
JAAC	*Journal of Aesthetics and Art Criticism*
Jaw.	"Jawāhir al-Sirr al-Munīr," attributed to Ibn Sabʿīn (MS. 7127 Ẓah.)
JRAS	*Journal of the Royal Asiatic Society*
JSAH	*Journal of the Society of Architectural Historians*
IJMES	*International Journal of Middle East Studies*
IQ	*Islamic Quarterly*
IS	*Islamic Studies*
Kahf	"Al-Kahf wa al-Raqīm fī Sharḥ Bism . . . ," al-Jīlī (MS. 4992 Ẓah.)
Lis.	*Lisān al-ʿArab al-Muḥīṭ*, Ibn Manẓūr.
Ras.	*Rasāʾil Ikhwān al-Ṣafāʾ wa Khillān al-Wafāʾ*, Dār Ṣādir ed.
Ris.	"Risāla fī Bayān ʿAdad al-Samāwāt wa al-Arḍīn . . . ," al-Āmidī (MS. 6102 Ẓah.)
SCR	*Studies in Comparative Religion*
Tuḥfa	"Al-Tuḥfa al-Mursala fī Waḥdat al-Wujūd," al-Hindī (MS. 5511 Ẓah.)
Yad	"Al-Yad al-Ajwad fī Istilām al-Ḥajar al-Aswad," al-Bīrāmī (MS. 7847 Ẓah.)
Ẓah.	Dār al-Kutub al-Ẓāhiriyya

Chapter 1

Discursive Order

Cosmology: An Overview

A study of cosmology and architecture from a premodern Sufi perspective presupposes some knowledge of the trends in Islamic cosmological thinking, of the sense in which cosmology can be related to architecture and of the terrains of mystical thoughts in which the relationship is grounded. This chapter presents a brief overview of cosmology in premodern Islam, a critical review of symbolism, being the prevailing method of reading cosmological ideas into architectural forms, and a reconstruction of a Sufi perspective on symbolism.

Cosmology is the science of the cosmos—its origin, structure, components, order, and governing laws. Its complex and multifaceted inquiry unfolds at the intersection of philosophy, theology, and natural sciences and is sustained by the human curiosity to know how we have come to exist and what happens to us when we die. Until the triumph of modern physics, cosmology was the prerogative of theologians, mystics, and philosophers, forming the core of religious sciences. All world religions provide their followers with a "logical" explanation of the creation, with a description of the cosmic landscape and order, and, most important, with a projection of what is awaiting them in the hereafter. Cosmological doctrines were thus significant not just for their scientific validity but also for the enforcement of moral and religious codes of conduct. Reward and punishment, the potent instruments of religious law and authority, can only work within a current cosmological system of popular appeal; hence the sensitive and volatile relationship religion has often had with science.

In Islam, the main sources of cosmological ideas were naturally the Quran and the *ḥadīth* (prophetic sayings). The Quran presents many references to cosmic elements—the Throne, the Footstool, the Pen, the Tablet, heaven, and earth—to the creation and resurrection, to paradise and hell, and so on, but mostly in an abstract way without weaving a complete and coherent cosmic picture. It is the *ḥadīth* that provide much of the information needed to piece the Quranic elements together into a coherent architecture.[1] In broad terms, two distinct modes of cosmological thinking can be traced in the Islamic tradition: theorized and untheorized. The untheorized mode was concerned with a collected body of statements made by the Prophet and his immediate companions, which provided, as it were, the nonnegotiable Islamic *truths*, the foundations necessary for cosmological reflections and speculations. The theorized mode was concerned with making sense of the Quranic and prophetic material and was cultivated in three different intellectual spheres: theology and polemics, philosophy and science, and hermeneutics and mysticism.

Related though they may be, the two modes of cosmological thinking developed, rather curiously, independent of each other. The untheorized mode, which appealed to mainstream religious authorities, formed part of the *ḥadīth* reporting science that was concerned with the authenticity of the statements and the credibility of the reporters. In this context, cosmological statements were transmitted and perpetuated in the *ḥadīth* books alongside statements concerned with daily matters, such as, prayer, ablution, marriage, divorce, pilgrimage, and so on. Early *ḥadīth* scholarship must have provided an effective way of appropriating and authenticating pre-Islamic cosmological conceptions and popular narratives that seemed to be in harmony with the new religion. Over the history of Islamic cosmological thinking, the *ḥadīth* corpus proved to be a powerful tool in the hands of clerics, who grew more and more suspicious and intolerant of "foreign sciences," until they prevailed in the sixteenth century when a *ḥadīth*-based genre of cosmological writing dominated over scientific and philosophic curiosity. Al-Suyūṭī's popular treatise *al-Hay'a al-Saniyya fī al-Hay'a al-Sunniyya*, which deals with what would have been perceived and presented as religiously "factual" and "authoritative" cosmological data, is a key text that represents this mode of cosmological thinking.[2]

The theorized modes flourished in the early periods, producing a rich spectrum of trends and ideas. With the *kalām* movement, described as Arabic scholasticism, which emerged in the ninth century, we have the early rationalists and polemicists who developed sophisticated cosmological arguments concerning such difficult issues as the existence of God, anthropomorphism, creation, nature of existence, free will, and determinism. This was led by the Muʿtazilites who were challenged and later succeeded by the Ḥanbalites and Ashʿarites. The *kalām* practitioners were theologians concerned with the understanding and interpretation of the divine revelation within rigorous linguistic context, taking the Islamic truths as the basis of their polemical engagements.[3] In this they differed from early philosopher-scientists, such as al-Kindī (d. c. 866), al-Rāzī (d. c. 925/935), and al-Bīrūnī (d. c. 1051), who were more inclined to start from observational knowledge and human reason in their cosmological thinking. And Ibn al-Rāwandī (d. c. 910), who attempted to forsake the religious truths altogether, points to the breadth of ideas that emerged in early Islam.

Within the intellectual sphere of philosophy and science Muslims made remarkable achievements in the fields of astronomy, mathematics, geometry, optics, geography, medicine, and alchemy. The new findings contributed significantly to the development and sophistication of their cosmological thinking. Persian, Indian, and other Near Eastern influences were absorbed into the Muslim worldview; however, it was Greek knowledge—a combination of Ptolemaic astronomy and Aristotelian philosophy and physics—that ultimately prevailed in the Islamic world. Philosophers and scientists maintained a rather precarious relationship with mainstream religious authorities, who were ready to attack whenever the Islamic dogma was being challenged. Al-Ghazālī's forceful attack on philosophy and Ibn Taymiyya's on Sufism and other schools are but two eminent examples.[4]

Motivated by spiritual fulfillment rather than scientific curiosity, the mystics cultivated a hermeneutical mode of cosmological thinking that wove together all aspects of available knowledge into a comprehensive whole. Mystics included Gnostics such as Ikhwān al-Ṣafā', al-Sijistānī, and al-ʿĀmilī and Sufis such as al-Ḥallāj, Rūmī, and al-Ghazālī. Islamic mysticism reached its zenith in the work of Ibn ʿArabī (d. 1240), whose multilayered and complex cosmology has since dominated the Islamic world. Unlike the philosopher-scientists who did not engage the *ḥadīth* literature, the Sufis elaborated the Quranic-prophetic model and integrated its terminology into their cosmological doctrines. In doing so, they provided a sophisticated, yet popular, framework for the *ḥadīth*-based cosmology, grounded in interwoven layers of hermeneutical interpretations that extended to every aspect of daily life. This, in a way and despite the tenuous relationship the Sufis had with mainstream Islam, assisted in the triumph of the al-Suyūṭī's unspeculative version of orthodox cosmology once the scientific thrust had abated. It is this trend of cosmological thinking that is explored in this study.

Underneath these different modes of cosmological thinking, there lay basic consistencies. The Platonic-Aristotelian duality of the *sensible* and the *intelligible*, the physical and metaphysical, along with the Ptolemaic geocentric model were uncontested. For over a millennium, from Mujāhid b. Jabr's (d. 722) very basic cosmography to Ḥaqqī's (d. 1780) most elaborate *Maʿrifat-nāme*, a remarkable consistency can be traced in the cosmic form and structure.[5] The Islamic cosmos consisted of the seen and unseen, the divine and human domains, with each having its own inhabitants, landscape, and order. The seen world was constructed of nine concentric spheres, seven planetary ones encompassed by the sphere of the fixed stars (the divine Footstool) and the utmost encircling sphere without stars (the divine Throne). The seven heavens rest on seven earths in the form of domed structures decreasing in size and positioned one within the other. As for the workings of the cosmos, it was seen to be regulated by a quaternary natural order of the four elements, mediated by many sets of four—four seasons, four natures, four humors, four directions, and so forth. Another consistency can be traced in the popular narrative of the creation that formed the starting point of many Islamic chronicles, integrated into many literary sources, and appeared in the seventeenth- and eighteenth-century Ottoman genre of architectural writings. *Risāle-i Miʿmāriyye*, the treatise on architecture, opens with a cosmological recount of the creation and structure of the world, followed by poetic reflections, whereas the *Selimiyye Risālesi* includes many references to the correspondence between a building and the cosmos.[6] Yet neither text provides theoretical articulation of the relationship between cosmology and architecture, nor indeed do other premodern Islamic sources. The discursive relationship was mainly the work of modern theorists working with the notion of symbolism.

Symbolism: A Critical Review

Symbolism is a current topic in many disciplines; its discourse is multifaceted. Psychologists, anthropologists, cultural theorists, social scientists, historians of religion, historians of art and architecture, philosophers and architectural theorists, have all developed viable dimensions of the discourse. The wide interest in the topic in the humanities and social science disciplines points to its significance in understanding humankind and its situations in the world. It is beyond the scope of this study to review all of these approaches, valuable though they may be. The perspective is narrowed down to one particular approach developed and promoted by a group of contemporary scholars, focusing mainly on medieval metaphysics and mystical sciences. The founder of this approach was the French metaphysician René Guénon (1887–1951), who devoted his life to the study and revival of traditional sciences, first through Hinduism and later through Islam. His ideas influenced two eminent scholars: Sri Lankan metaphysician and art historian Ananda Coomaraswamy (1877–1947) and Swiss metaphysician Frithjof Schuon (1907–1998). The distinct approach to the study of symbolism and metaphysics these scholars developed attracted many followers, who contributed to their compelling theoretical analyses and shared their passion for tradition.

Driven by a great passion for medieval metaphysics, these "traditionalists" or "perennialists" (from "perennial philosophy"), if I may so describe them, sought to reintroduce into the life of modern Western society the badly needed "spiritual essence" and long forgotten "traditional wisdom." Spirituality and traditional wisdom, seen as direct manifestations of a collective preoccupation with divine revelations, were seen as the indispensable, priceless possessions of the East the industrialized West had lost. It might have advanced in natural sciences, however, since the European Renaissance, they argue, Western civilization has witnessed but a steady decline in the intellectual, spiritual, and artistic aspects of life, culminating in the "dark age" of modernism. Such a decline would not have been a matter of great concern, they add, had it not infectiously spread fast and wide, mindlessly destroying the Eastern traditions in the name of modernity and progress. "Cain who killed his brother Abel, the herdsman, and built himself a city," Coomaraswamy writes, "prefigures modern civilization, one that has been described from within as 'a murderous machine, with no conscience and no ideals,' 'neither human nor normal nor Christian,' and in fact 'an anomaly, not to say a monstrosity.'"[7]

Among the first circle of scholars who followed and contributed to the founding of the perennialist approach were Titus Burckhardt, Seyyed Hossein Nasr, and Martin Lings. These are of a special significance to this study, for they have extended the perennialist approach to the study of premodern Islam and its rich artistic and architectural heritage. Their works have opened up a new horizon of understanding and provided new intellectual tools for rethinking the field's long-established art history paradigms. Driven by a keen interest in Sufi spirituality, they sought to show how in premodern Islam the mystical experience was inextricably linked to most aspects of life, including the creative acts of making. Sufi spirituality, they argued, can be traced in various modes of expression, the most conspicuous of which are art and architecture. In the perennialist project, Sufism becomes more than an object of academic interest: it becomes a way of life and a pursuit for knowledge, hence, the great passion and conviction with which they write.

The perennialist approach has inspired several studies on traditional art and architecture by figures such as Ardalan, Bakhtiar, Critchlow, and Snodgrass.[8] Along with the studies of the founding figures, these sources have articulated "symbolism" as a distinct approach to the study of traditional built environment. Their efforts were complemented by the writings of such insightful and prolific scholars as Annemarie Schimmel, Henry Corbin, Louis Massignon, Toshihiko Izutsu, and Hellmut Ritter, who made significant contributions to the interpretation and understanding of medieval mystical sciences in general and that of Sufism in particular. The approach of symbolism has also benefited extensively from the work of the eminent Romanian anthropologist and historian of religion Mircea Eliade. Eliade wrote from outside the perennialists circle; however, his studies were instrumental in refining the methodological tools of symbolism. Apart from expounding the methodological grounds of religious symbolism, Eliade articulated a theoretical framework for interpreting traditional mythology and understanding its social and religious functions. This has strengthened the method of symbolism by enabling "myth" to play an effective mediatory role between abstract scriptural principles and concrete human experiences. Eliade wrote very little on Islam and Islamic mythology; however, his comparative method and ecumenical perspective underpin the pertinence and usefulness of his analyses.

The writings of the perennialists provided a fertile ground for theoretical reflections on premodern art and architecture across cultural boundaries and outside the conventional bounds of art history. Central to their approaches is the notion of symbolism, viewed as the core of spiritual engagement and the language of premodern artistic expressions. The perennialists' project, however, involves other foundational themes, which are introduced briefly here.

The Transcendent Unity of Religions

The perennialists viewed religion and tradition to be the main elements characterizing premodern society, be it Islamic, Christian, Hebrew, Buddhist, or Hindu. It is religion, they argue, that provides a community with the immutable divine principles that govern the worldview of its members, and it is tradition that weaves these principles into peoples' modes of living, thinking, and making, while being handed down from one generation to another. Religion, the source of heavenly truths without an awareness of which humanity would eventually bring about its own destruction, is presented as having two related aspects: exoteric and esoteric. The exoteric aspect *reveals* the diverse, the different, whereas the esoteric *contains* the same, the essential. It is the unifying core, the summit to which all paths lead and at which they all converge. "In the spiritual world and, still more, the universal order," Guénon writes, "it is unity that presides at the summit of the hierarchy."[9] The mountain metaphor is frequently used to illustrate the shared conviction that all religions are but so many roads that lead to one and the same summit. In "Paths That Lead to the Same Summit," Coomaraswamy takes an ethical stance to argue that all great religions of the world have valid claims to truth, which must be respected and understood in a comparative mode and not just tolerated.[10] Doctrinal differences, conspicuous at the mountain's wide base, should not prevent us from recognizing the inner meanings or seeing the *peak* where all differences vanish. Such recognition does not necessarily demand a change in the path that one finds oneself on naturally, for "he who goes round about the mountain looking for another is not climbing."[11] Asserting the necessity of this approach for cross-religious understanding, Coomaraswamy writes: "The greatest of modern Indian saints actually practiced Christian and Islamic disciplines, that is, worshiped Christ and Allah, and found that all lead to the same goal: he could speak from experience of the equal validity of all these 'ways,' and feel the same respect for each, while still preferring for himself the one to which his whole being was naturally attuned by nativity, temperament and training."[12]

The Transcendent Unity of Religions was the title of Schuon's first book, in which he articulated the theme from a metaphysical perspective.[13] From the outset, Schuon was careful to clarify the differences among metaphysics, theology, and philosophy, in order to assert that the doctrine of the unity of religions expresses a metaphysical reality, hence its transcendence, and not just a philosophical or theological argument. To illustrate the difference between metaphysical and theological knowledge, and indirectly to emphasize the unity of religions, Schuon uses the metaphor of light. He compares metaphysical knowledge to the awareness of the "colorless essence of light and its character of pure luminosity" and theological knowledge to the assertion of light's particular colors. Although both modes overlap in their distinction between light and darkness, they differ in the level of reality they disclose. Metaphysics discloses universal truths, whereas theology discloses divine revelations, which are but particular expressions of the universal truths. Philosophy is further removed from theology in that it deals with rational concepts. Religions, thus viewed, translate universal truths into dogmatic languages that are accessible by the wider public through faith. But the colorful variations of the dogmatic beliefs fade at the level of universal truths where all religious differences disappear. In brief, the reality of the transcendent unity of religions is shown to reside in the oneness of the Truth that governs all modes of manifestation and existence, and in the oneness of the human race that alone has the capacity of tracing the process of differentiation back to its transcendental source.[14]

Perennial Philosophy

If religions are so many expressions of one and the same primordial Truth, it is then only natural for them to share a universal focus and a mode of expression whereby aspects of the Truth can be accessed, comprehended, and communicated. This focus forms the subject of what in the Latin tradition is called "*philosophia perennis*," "perennial philosophy," whose primary mode of expression is shown to be traditional symbolism. Perennial philosophy, also referred to as "universal" and "eternal" philosophy, differs from modern philosophy in that it is not a system of thought or theories about the nature of the world, but as Coomaraswamy puts it, "a consistent doctrine." As such, perennial philosophy is not concerned with historicized realities, that is, "with conditioned and quantitative experience, but with universal possibility."[15] It is described as "perennial," "eternal," and "universal" because of the immutability of its focus, eloquently defined by St. Augustine as "Wisdom uncreate, the same now as it ever was, and the same to be for evermore."[16] Thus understood, perennial philosophy is the vehicle that reveals "the concordances between all traditional forms," which may be seen as projecting "genuine synonymies."[17] In his preface to Schuon's *Islam and the Perennial Philosophy*, Nasr captures the crux of this doctrine in saying: "The *philosophia perennis* has come to signify for those devoted to traditional studies an eternal truth at the heart of all traditions."[18]

By definition, perennial philosophy presupposes a universal perspective, one that demands cross-religious understanding even if one's concern is restricted to a particular tradition. For the ultimate aim is to allow the particular to be seen and understood in light of the universal and not only in its own sociohistorical context. Where the particular bares no relation to the universal it is considered to be insignificant, since it does not signify anything beyond its particularity. Moving across religious and doctrinal boundaries, however, requires insight into the universal principles that legitimate such movement, without which one runs the risk of syncretism. Guénon sees syncretism as symptomatic of a condition where universal truths are no longer in perspective and "can be recognized whenever one finds elements borrowed from different traditional forms and assembled together without any awareness that there is only one single doctrine, of which these forms are so many different expressions, or so many adaptations to particular conditions related to given circumstances of time and place."[19]

The universal perspective of perennial philosophy naturally deplores today's wide emphasis on relativity and the historicity of human experience and runs counter to the prevailing views that all truths, including the religious ones, are simply the outcome of certain modes of representation and cultural construction. In *Logic and Transcendence* Schuon argues against relativism on the basis that such views hinge on an implicit assertion that is negated by the concept of 'relativism' itself.

> Relativism sets out to reduce every element of absoluteness to a relativity, while making a quite illogical exception in favor of this reduction itself. In effect, relativism consists in declaring it to be true that there is no such thing as truth, or in declaring it to be absolutely true that nothing but the relatively true exists; one might just as well say that language does not exist, or write that there is no such thing as writing . . . The assertion nullifies itself if it is true, and by nullifying itself logically proves thereby that it is false; its initial absurdity lies in the implicit claim to be unique in escaping, as if by enchantment, from a relativity that is declared alone to be possible.[20]

Schuon concludes this book with a discussion on certainty, showing the crucial function of perennial philosophy in a human life "studded with uncertainties" and not just as an academic pursuit. In everyday life, he argues, one is presented with numerous possibilities to choose from and is confronted with many events about which to make decisions. The universal divine truths that concern human existence in this world as well as in the hereafter, with which perennial philosophy is concerned, provide consistent criteria against which one's choices and decisions can be measured. Without the constant presence of such absolute criteria one would ultimately be lost in the world of uncertainties and be drowned in the ocean of possibility. Perennial philosophy thus calls for perpetual renewal of the understanding of universal truths so that they retain their relevance to people's lives. Although it has no history with regard to the realities it appeals to, perennial philosophy has maintained a historical presence through the spiritual authorities who have devoted their lives to expounding its principles in various times and places. This is viewed to have been the case throughout the world until the advent of European modernism.

Modernism and Spiritual Decline

European modernism, founded on humanism and a preoccupation with aesthetics and profane sciences, the perennialists argue, has shifted the focus from God to man, thus breaking the continuity of tradition, eclipsing the spiritual pursuit, and consequently, marking the beginning of the decline of human civilization. This is a central theme in almost all of their writings, backed by a relentless attack on many aspects of modern ideals, sciences, and most important here, modes of artistic production. Guénon devoted much effort to initiate such a systematic attack, which he presented in *East and West*, *The Crisis of the Modern World*, and *The Reign of Quantity and the Signs of the Times*. Both Coomaraswamy and Schuon maintained in almost all of their works the intensity of Guénon's criticism, taking every opportunity to remind their readers of how psychologically corrupt, intellectually deranged, and spiritually bankrupt modern life has become.

This ideological position has led to a sharp distinction between modern and premodern modes of thinking and making, which has theoretical and methodological implications. The perennialists present the distinction between modern and premodern as synonymous with that of West and East and of antitraditional and traditional. Guénon passionately argues that the rift dividing East from West, as "one of the most noticeable features of the modern world," has emerged from the antitraditional sentiment and associated mentality promoted by the "modern West."[21] So long as there were "traditional civilizations," that is, peoples in tune with tradition as in premodern times, Guénon argues, "no ground for a radical opposition between East and West existed."[22] But when the West took the turn toward humanism and profane sciences it began to distance itself from the East, which remained steadfast on the traditional path. Thus, a multifaceted opposition between two geo-mentalities was constructed, forming the basis for a range of binaries, such as knowledge and action, sacred and profane, rational and intuitive, egocentric and anonymous, and so on. In *Light on the Ancient Worlds*, Schuon explains this opposition by reference to what preoccupies each of the modern and traditional sciences. He writes: "Modern science, which is rationalist as to its subject and materialist as to its object, can describe our situation physically and approximately, but it can tell us nothing about extra-spatial situation in the total and real Universe . . . Profane science, in seeking to pierce to its depths the mystery of the things that contain—space, time, matter, energy—forgets the mystery of the things that are contained."[23]

Traditional science, by contrast, is viewed to be preoccupied with metaphysical realities, with the *mysteries* that are contained. Preoccupation with metaphysical truths, the object of perennial philosophy, and the applicability of these truths across different religious contexts formed the basis of the perennialist discourse on "traditional" art and architecture. In this discourse the adjectives *traditional, medieval, oriental,* and even *true* were all used synonymously to define a condition that is distinct from that of the modern West. The sharp distinction between modernity and tradition was drawn on the basis of religious efficacy and the presence of spirituality. In medieval theocentric societies when religion was effective, the perennialists maintain, spirituality was intensely present in people's lives and clearly visible in their modes of thinking and making. Modernism, ushered in by the so-called European Renaissance, has introduced new living conditions and modes of thinking that led to systematic erosion of religious values. Against the millennial presence of tradition, however, the brief history of modernity can only appear as a peculiar abnormality. "We are peculiar people," Coomaraswamy writes, "I say this with reference to the fact that whereas almost all other peoples have called their theory of art or expression a 'rhetoric' and have thought of art as a kind of knowledge, we have invented an 'aesthetic' and think of art as a kind of feeling."[24] In this context, studying medieval or traditional art and architecture requires not only an awareness of the changes modernity has introduced but also the use of an appropriate approach to uncover their spiritual content and to facilitate a proper understanding of their meanings. The approach is, of course, that of perennial philosophy mediated by traditional symbolism. In his forward to Ardalan and Bakhtiar's *The Sense of Unity,* Nasr summed up the ultimate aim of the perennialist project: "There is nothing more timely today than that truth which is timeless, than the message that comes from tradition and is relevant at all times. Such a message belongs to a now which has been, is, and will ever be present. To speak of tradition is to speak of immutable principles of heavenly origin and of their application to different moments of time and space."[25]

The Necessity of Symbolism

Dealing with the language of the timeless Truth and the immutable principles of tradition requires an approach that is in tune with the traditional views. As the medium of traditional artistic expression and the language of *philosophia perennis,* symbolism, the perennialists argue, is the most appropriate approach for comprehending the inner meanings of traditional art and architecture and for penetrating deep into their worlds of spirituality and metaphysics. Symbolism is presented as the "language" of religion that divinity "speaks," using allegories and similitude. Being ontological in nature, the language of symbolism communicates the fundamental and universal conditions of existence. It is, as Martin Lings puts it, at once "the most important thing in existence" and "the sole explanation of existence."[26]

Symbols and signs, be they verbal or visual, are commonly understood as means of communication. The lexical definitions refer to "sign" and "symbol," in the general sense, as "something used or regarded as standing for or representing something else."[27] The perennialists, however, distinguish between signs and symbols in terms of their referents. "The references of symbols," Coomaraswamy says, "are to ideas and those of signs to things."[28] Symbols refer to immaterial concepts—it is "the representation of reality on a certain level of reference by a corresponding reality on another"[29]—whereas signs refer to material objects that stand on the same level of reference. One object can be both a symbol and a sign according to its referent: "[T]he cross, for example, is a symbol when it represents the structure of the universe, but a sign when it stands for crossroads."[30]

The main premise upon which hinges the notion of symbolism is that material objects, tectonic or otherwise, are capable of embodying abstract concepts that lie beyond the confines of their materiality. This basic understanding assumed in the Western tradition a decisive philosophical formality in the Platonic distinction between the *sensible* and the *intelligible*. Medieval Muslim scholars, who inherited and developed the intellectual tools of Greek philosophy, maintained this through the divide between *al-ḥissī* and *al-ʿaqlī*. This was, in a sense, legitimized by the Quranic polarity of the *seen* and the *unseen*. Accordingly, symbolism is understood to be based on the correspondence between these two domains of reality: the inferior reflecting the superior, the visible materializing the invisible, and the physical representing to the spiritual. Predicated on the universal perspective of *philosophia perennis*, symbolism becomes an ontological inquiry, an inquest into the hierarchy of being, and an intellectual journey to the inner worlds of universal realities. Thus viewed, symbolism often becomes a pursuit of esoteric knowledge, reaching far beyond the mere visual objectification of religious or cultural values. In the perennialist project, architectural symbolism is developed on these conceptual grounds. Forming an integral part of the sensible world, architectural forms are considered as eminently appropriate to act as symbols.

Symbols, the perennialists explain, are of two fundamentally different kinds: universal or natural and particular or conventional. Universal symbols are those whose symbolic significance derives from their innate nature, such as geometrical or numerical symbols, whereas particular symbols are those whose symbolic significance relates to a particular tradition. By virtue of their very nature, universal symbols are regarded as primordial and transcultural, whereas particular symbols, which include particular interpretations of universal symbols, vary in different traditions. Particular symbols, also described as arbitrary and accidental, may be sanctified by human or divine intervention that make them the loci of transcendental meanings, such as the cross as a symbol of resurrection in Christianity and alphabetical symbolism in Islam. They are empowered by a communal acceptance and participation in their spiritual significance.

Universal symbols are ontologically linked to, and determined by, their referents. Hence, symbols of the infinite and the timeless have the capacity of revealing aspects of the infinite and the timeless itself.[31] Symbols express the quality of the infinite with respect to their finitude, finding through such expression their transcendental dimensions, their opening beyond the finitude of their own realm. By mediating ultimate reality through things or actions, symbols receive the quality of the higher realities they mediate, enabling us to comprehend them with respect to the finitude of our own existence. In the experience of sacred places, for example, symbols of the sacred reveal something of the "Sacred-itself" and produce the experience of sacredness in those who are experiencing them. They are thus capable of revealing a modality of the real, the sacred, or the absolute and of unveiling the deep and profound structure of the universe. They form the "alphabet" of the universal language of religion whereby ultimate reality expresses itself, revealing a coherent picture of existence and of the world. In this sense, symbols require no justification; the only measure of their validity is their adequacy to the higher realities they express.

Symbolism adds meanings to objects and practices without affecting their proper and immediate value. Here meanings are not seen as being "'read into' symbols or added to them as a conceptual garnish. On the contrary, they are deemed to inhere within the form of the symbol in a manner analogous to that in which natural law inheres within physical phenomena, or as mathematical principles reside in the very nature of numerical or geometrical phenomena."[32] In being so, the meanings of symbols are not intentionally constructed but rather discovered or revealed through reflections on transcendental realities, and consequently the efficacy of a symbol does not depend on its being understood. "A religious symbol," Eliade writes, "conveys its message even if it is no longer *consciously* understood in every part. For a symbol speaks to the whole human being and not only to the intelligence."[33]

Symbols are multivalent. They can simultaneously express a number of meanings "whose continuity is not evident on the plane of immediate experience."[34] One symbol may refer to a plurality of contexts, and its significance can be operative at a number of different levels. To fully explain the meanings of a symbol, Eliade argues, is not a simple task, and to exhaust the significations of those concerned with divinity is not possible. Confining a symbol to only one of its significations as the most important is therefore reductive. For the significance of a symbol lies in revealing the unity and continuity between the different levels it reveals. Tracing the various expressions of a symbol reveals the density of meanings and resonance across multiple contexts. In becoming a symbol, a concrete object is thus enriched with layers upon layers of interrelated meanings, which are not necessarily evident through an immediate experience. "The interdependence of the valences of a symbol and the homology of its different contexts," Eliade explains, "ought not be understood as a monotonous repetition of the same message on different levels . . . Each context of a symbol reveals *something more* which was only unformed and allusive in the neighbouring contexts."[35] In this sense, symbols imbue human existence with significance by pointing to a more profound, more mysterious side of life, to the miraculous and "sacramental dimensions of human existence."[36] For this reason, symbols related to ultimate reality are viewed as a source of inspiration and revelation.

In summary, the perennialists approach the question of artistic production from the viewpoint of creative imagination and religious inspiration. They focus primarily on the ideas, rituals, and cosmology within the matrices of which an artefact is produced, rather than the historico-cultural conditions that facilitate such production. Through symbolism they establish a continuity among the human, cosmic, and divine modes of being, providing a means to interpret the human conditions of existence in cosmological terms. In their perspective, "symbolic thought makes the immediate reality 'shine'" by enabling us to see human makings through a cosmological frame, wherein "everything holds together in a closed system of correspondences and assimilations."[37]

History and Symbolism

The perennialist discourse is of course not without problems and critics. Historians of Islamic art and architecture have already discredited it for its essentialist tendency and lack of methodological rigor, yet their critique has largely been concerned with its perspective on art and architecture. The perennialists, as we have seen, have adopted a hermeneutical approach in their attempts to reinstate an appropriate interpretive context for understanding the metaphysical foundation of symbolism in traditional art and architecture. They also strove to uncover the mystical and cosmological contexts of human makings, the dimensions that are almost completely lost in contemporary discourses and practices. Their passion reveals a strong sense of conviction and genuine desire for spiritual fulfillment. Academic curiosity is clearly not their prime motivation: their endeavor is above all a search for truth. No critical stance has, therefore, been projected in regard to the approach of symbolism. In fairness, however, one should not expect to encounter such criticality, for that would undermine the very certainty of the doctrine of symbolism itself and the *philosophia perennis* upon which it is predicated. A degree of anonymity is also emphasized, which reflects the sense of universality their discourse promotes. "It is not the personal view of anyone that I shall try to explain," Coomaraswamy affirms, "but that doctrine of art which is intrinsic to the *Philosophia Perennis*."[38] An antimodern sentiment motivates the main line of criticality pursued, which, although insightful, seems more ideological than methodological in focus.

Reflecting on the modern approaches to the study of premodern art and architecture, the perennialists have criticized many aspects of modern sciences, which they dismiss for adopting predominantly a reductive historical perspective that is unsuitable for dealing with spirituality and metaphysics. "The history of art, being a modern science," Burckhardt argues, "inevitably approaches Islamic art in the purely analytical way of all modern sciences, by dissection and reduction to historical circumstances. Whatever is timeless in an art—and sacred art like that of Islam always contains a timeless element—will be left out by such a method."[39] The perennialists maintain that historically and culturally contextualized studies, valuable though they may be, remain incapable of engaging the mystical and metaphysical dimensions of the artistic production. "A form, though limited and consequently subject to time," Burckhardt adds, "may convey something timeless and in this respect escapes historical conditions, not only in its genesis—which partly belongs to a spiritual dimension—but also in its preservation, to a certain extent at least."[40]

The historians remain unconvinced, however. In *The Topkapi Scroll—Geometry and Ornament in Islamic Architecture,* Gülru Necipoğlu presents an insightful critique of the traditionalist discourse, focusing mainly on studies dealing with geometry and ornaments. Her main concerns center on the following points: the uncritical application of the theological concept of '*tawḥīd*' (divine unity) to emphasize the unity of Islamic art and architecture; the lack of contextualized and historicized analyses; the dogmatic and essentialist approach to the subject; and the "sweeping generalizations unsubstantiated by concrete data."[41] Necipoğlu also points out where the perennialists have continued to operate within the orientalist and nineteenth-century conceptual paradigms, despite their new antiorientalist, antimodernist guise. Earlier, W. K. Chorbachi in a lengthy article presented a similar critique.[42] Speaking from an empirical, scientific standpoint, Chorbachi sought to dispel the confusion that beset the field of Islamic geometric pattern because of the lack of a common language. Her main concern is the extension of the mystical doctrine of the Unity of Being *(waḥdat al-wujūd)* to the field of geometry, by proposing that all geometric articulations in Islamic art and architecture can be reduced to the division and subdivision of a circle, conceived and presented as the symbol of divine unity. "On occasion," Chorbachi writes, this view "is pushed to the point of scientific fallacy."[43] She seems rather puzzled with how simple geometric exercises in rotational symmetry or the interlocking eight-pointed star and cross pattern can become symbols of divine presence and absence or the Breath of the Compassionate. "I wonder," she wrote, "if the artisan who made this design thought of it as form, expansion, contraction and the Breath of the Compassionate God?"[44] Chorbachi points out that "to believe in mysticism and to follow in its practices and experience its positive effects" is one thing, but to promote new mystical interpretations "under the guise of historical truth," especially when no documented evidence is given, is quite another.[45]

Chorbachi's and Necipoğlu's critiques reflect the stance of the influential Islamic art and architecture historian Oleg Grabar, who initiated and sustained an insightful critique of symbolism and the perennialist approach. In "Symbols and Signs in Islamic Architecture," Grabar discusses the methodological problems of the current discourses on symbolism, identifying three inherent shortcomings. First is the "absence of scientific precision," that is, the lack of an explicit link between historical data and mystical interpretations of built forms. Second is the ambiguity of the Islamic character that is considered unique but "never described." And third is the absence of the contemporary context that grounds interpretations in existing literary documents, which "would prevent the unavoidable impression of modern constructs, perhaps valid to modern man, applied to traditional forms."[46] Grabar also points out that the universal views of the perennialists "owe little to broad symbolic theories" and warns against Eliade's approach through which "unique cultural experiences can much too easily be transformed into meaningless and obvious generalities."[47] While pointing to many pertinent areas requiring further development, Grabar provides valuable insights into how to approach and think about this complex topic. In a later series of lectures published as *The Mediation of Ornament,* however, Grabar seems more inclined to abandon the idea of inherent meaning altogether and, consequently, to dismiss the viability of the approach of symbolism in favor of a psychoanalytical approach grounded in the nature of visual perception. In his conclusive remarks, Grabar explains that "works of art and in general the visually perceived environment have an extraordinary power in shaping the lives and thoughts of men and women" and that "ornament is the ultimate mediator, paradoxically questioning the value of meanings by channelling them into pleasure."[48] This was the stance from which he later analyzed the Dome of the Rock in *The Shape of the Holy*. Grabar's views seem to have influenced Doris Behrens-Abouseif's approach in her *Beauty in the Arabic Culture*, wherein she argues that the norms of beauty in the Arab-Islamic culture were autonomous, pleasure-oriented, and independent of moral and religious criteria. Although her proposition is predicated on the lack of contrary evidence and loosely argued, her view contrasts that of the perennialists who would ardently disagree that meanings in Islamic art and architecture are a matter of aesthetics and psychology instead of symbolism and epistemology.

While recognizing the value and validity of the art historical critique, it is important to acknowledge the inherent dissimilarity between the perennialist and art historical approaches that cannot be adequately and fairly measured against one set of objective principles of academic research. A fundamental difference between the two approaches lies in the purpose and usefulness of each undertaking. Following the well-established methods of historical research, Islamic art and architecture historians have concerned themselves with the *accurate* and *objective* reconstruction of the historical reality. Their domain of influence and the usefulness of their discourse are confined mainly to specialized academic circles. The perennialists, by contrast, strive to *re-present* the historical reality in ways that serve the dogmatic belief of the Muslim community. In pursuit of truth, their research communicates matters of personal belief, and as such it tends to have a popular appeal. The perennialists approach the past for *useful renewal* of a perennial wisdom, seeking engagement, self-edification, and spiritual fulfillment, whereas art historians approach the past for *objective reconstruction* of the historical reality, seeking to satisfy an academic curiosity and a desire for knowledge. Clearly, this significant difference extends beyond methodology, making it difficult to measure the rigor of both approaches against the same criteria.

The art history critique, valuable as it may be, has remained largely confined to the broad methodological problems of the perennialist discourse, without tackling the approach of symbolism from within and on its own ground. From the art history perspective, it seems sufficient to point out the lack of rigor and of tangible evidence, textual and contextual, that supports the perennialists' claims to dismiss the validity and usefulness of their approach. Interestingly, this is exactly the same tactic used by the perennialists to dismiss the approach of art historians. Ironically however, the debate between the perennialists and art historians, despite the profound dissimilarity of their ideological and methodological positions, converges at one crucial point: the search for the "Islamic." The question that lies at the heart of both approaches seems to be: what is the legitimate mode of capturing this all-unifying adjective, history or symbolism, culture or spirituality? In their attempts to explain the *difference* that distinguishes art and architecture of premodern Islam, both the perennialists and the art historians reveal a continuous struggle with this deep-seated orientalist preoccupation. Both seem preoccupied with articulating, in one way or another, a coherent and consistent discourse to explain the *Islamic difference*. But whereas the perennialists passionately argue for the underlying *unity* of difference, art historians, who ironically were first to propose this unity in the nineteenth century, have recently been fervently arguing for its *diversity*.

Notwithstanding the methodological problems of the perennialist approach, the value of its discourse remains in the theoretical possibility it affords and the horizon of thinking it opens up, both of which are remote, if not alien, to the sociocultural concerns of art history. Its merit still lies in enabling the study of art and architecture to become a cosmological inquiry, an intellectual inquest into the hierarchy of being, and a philosophical journey into the worlds of universal ideas. These are critically pursued, explored, and examined here within the premodern Islamic literary context but without engaging with the perennialists' ideological package or with their preoccupation with the Islamic. The main focus here is the reconstruction of the premodern spatial sensibility that locates the meanings of human making within a complex web of cosmological correspondences. The search for the Islamic is considered to be a different task that is made all the more problematic by the notion of spatial sensibility.

Spatial sensibility, as articulated here, is concerned more with the difference between modern and premodern modes of spatial ordering rather than with that between the Islamic and the non-Islamic, although the focus is on the Islamic perspective. While the analyses seek to illustrate the coherence and consistency of this spatial sensibility in premodern Islam as it discloses its workings in a variety of contexts, the study reveals, albeit indirectly, a fundamental discontinuity between the modern and premodern conceptions and practices. Architectural symbolism, as articulated in both the art historical and the perennialist discourses, disregards the spatial sensibility both of the interpreter and of those whose work is being interpreted. They are, in other words, indifferent to the *spatiality of difference*. In being so, the method of symbolism tends to free the interpreter form a crucial constraint while imbuing its own theoretical tools with a sense of universal applicability. It also establishes an implicit continuity between its retrospective and projective modes of analysis, allowing historical interpretations to easily slip into some forms of design theories. In revealing the cosmological "terrains" of the premodern Islamic sense of spatiality, this study argues that in our modern conditions we can only access the forms in which this sense manifest, but not the sense itself. The premodern spatial sensibility has been irreversibly changed: it can neither be revived nor repossessed.

Theoretical Distancing

Whether the focus is historico-cultural or mythico-spiritual, my approach emphasizes, our understanding of premodern symbolism remains a modern engagement. Modern conditions have introduced a theoretical distance between the symbol and its referents that has irreversibly altered its efficacy. What in a premodern context used to be intuitively available has now become the object of discursive understanding. It is, therefore, important to stress that the barrier of consciousness that is commonly recognized today as separating modern subjects from their traditions must also be seen as distancing them from the immediacy of symbolism. The perennialists overlook the fact that constructing layers of theoretical intermediaries between myth and architecture and between an object and its referents is a modern necessity. As David Kolb observes in *Postmodern Sophistications*, our ability to talk about tradition as a worldview with its own logic that is distinct from an objective world and from our subjective experience of this world, is the result of a new modern condition.[49] It is not surprising, therefore, to find no literature on architectural symbolism in medieval Islam. In fact, the whole discourse of symbolism, as constructed in modern sources, is alien to premodern Islamic literature, notwithstanding the presence of the mode of thinking that seems to support it. Yet the presence of a mode of thinking mediated by a unique spatial sensibility is one thing, whereas reconstructing the theoretical context within which this mode is supposed to have operated is quite another. It is in this theoretical reconstruction where the alienness emerges. It is our current intellectual conditions that demand such a theoretical construction to explain the association of the abstract meanings, already alienated, with concrete objects, an association that we have come to define as "symbolic." The discourse of symbolism, one may argue, is no more alien to premodern Islam than that of art history: both can be described as fictitious reconstruction of the past driven by modern preoccupations.

Acknowledging the inevitability of Kolb's three worlds scenario, this study engages tradition in a different way to that of the perennialists and art historians. Focusing on the specificity of the religious experience, the study accentuates the interpretive distance between the modern subject and the premodern object. It foregrounds the distinct spatial sensibility of the premodern in order to highlight the implicit discontinuity and disjunction between the retrospective (historical readings) and the projective (design theories) representations of difference. While exploring and revealing aspects of the Islamic difference, this approach highlights the impossibility of repossessing and recreating difference by careful manipulation of selected formal and spatial vocabulary.[50] The spatiality of difference is not reproducible by the appropriation and manipulation of architectural forms.

Sufism

Sufism *(taṣawwuf)* is an Islamic phenomenon associated with piety, ascetic life, and spirituality that emerged in the early formative period. Some scholars trace it back to the Prophet and his immediate companions, such as Abū Dhurr al-Ghaffārī (d. 651), Abū al-Dardā᾽ (d. 652), and al-Khuzāʿī (d. 672), however, the first commonly recognized Sufi personality is Ḥasan al-Baṣrī (d. 728).[51] Ibn Khaldūn (d. 1406), the celebrated Andalusian historian, regards *Sufism* and *Islam* as two synonymous terms. The piety and devotion of the first generation of Muslims provided, in his view, the model for Sufism. It was only later, and because of the inevitable distancing from the spirit of the religion with the march of time that Sufis, who remained attached to the spirit of Islam, came to be known and distinguished from others.[52]

Foundation

Most references on Sufism begin with a discussion of the name because it is shrouded with ambiguity with regard to its origin and meaning. Linguistically, it derives from *ṣūf*, "wool," and hence *ṣūfī* literally means "woollen," referring by extension to "one who wears wool." This was the view of al-Sarrāj (d. 988), who associates the term with the Sufis because of the woollen garment they used to wear.[53] Al-Qushayrī (d. 1074) disagrees on the ground that the wearing of a woollen garment, though familiar among them, was not one of their consistent practices.[54] Some relate the name to *ṣafāʾ*, "purity," to *ṣafwa*, "elect" or "elite," others to *ṣuffa*, as in *ahl al-ṣuffa*, the ascetics the Prophet used to shelter in a "shaded place" *(ṣuffa)* in his house.[55] Al-Qushayrī argues that none of these terms bears linguistic affinity to the name. Martin Lings proposes that the name could have been "first aptly applied to a small group who did wear wool and that it was then indiscriminately extended to all the mystics of the community in order to fill a void; for they had as yet no name, and since they were becoming a more and more distinct class, it was becoming more and more necessary to be able to refer to them."[56]

Since its emergence Sufism has continued to play a significant role in the intellectual, sociocultural, and political life of Muslim communities until its rapid decline in the nineteenth and twentieth centuries. Sufism might have started as isolated ascetic communities, but it rapidly gained popularity, spreading all over the Islamic world. Sufism is recognized to have been the main agency through which Islam spread into East and Southeast Asia and central Africa. In "The Mystic Path" Fritz Meier charts the historical development of Sufism in four stages: preclassical Sufism (eighth century), classical Sufism (ninth–tenth century), postclassical Sufism (eleventh–thirteenth century), and neoclassical tendencies (thirteenth–fourteenth century).[57] A classical phase may be considered with regards to the establishment of the phenomenon's identity and basic religious techniques through such legendary personalities as Dhū al-Nūn al-Miṣrī (d. 861) in Egypt; al-Muḥāsibī (d. 857), Abū Saʿīd al-Kharrāz (d. c. 899), al-Junayd (d. 910) and Ibn ʿAṭāʾ (d. 922) in Mesopotamia; and Abū Yazīd al-Bisṭāmī (d. 874/877–8), Abū Ḥafṣ al-Ḥaddād (d. c. 874), and Abū Bakr al-Wāsiṭī (d. c. 932) in Iran. However, as far as doctrinal development is concerned the period between the twelfth and sixteenth centuries is the golden age of Sufism.[58] It witnessed the maturity of the Sufi sciences after the systemization of the tradition by figures such as al-Sarrāj (d. 988), Abū Ṭālib al-Makkī (d. 998), al-Kalabādhī (d. 990/4), al-Sulamī (d. 1021), al-Qushayrī (d. 1072), and al-Hujwīrī (d. 1072–7).[59] The following generations included the most eminent and influential masters, such as al-Ghazālī (d. 1111), al-Jīlānī (d. 1166), ʿAṭṭār (d. 1190/c. 1220), al-Suhrawardī (d. 1234), Ibn al-Fāriḍ (d. 1235), Ibn ʿArabī (d. 1240), Rūmī (d. 1273), al-Shādhilī (d. 1258), al-Qāshānī (d. 1329), Naqshband (d. 1389), al-Jīlī (d. 1428), and Jāmī (d. 1492), through whom Sufism reached its zenith. While some of these great Sufi masters, such as al-Jīlānī, al-Shādhilī, and Naqshband, established mystical orders *(ṭarīqa)* that were to integrate Sufism into the society and to spread and perpetuate Sufi teachings throughout the Islamic world up to the present time, others, such as Ibn ʿArabī, Rūmī, and al-Jīlī, produced a wealth of religious and mystical literature and profound poetry that were to shape the intellectual life of Muslim communities until the dawn of the twentieth century.[60]

The contributions of these and many other formidable Sufi thinkers to various aspects of the religious and intellectual sciences in premodern Islam have shaped the Muslim worldview and underpinned its modes of thinking. In the conclusive remarks to his recent book on the history of Islamic mysticism, Alex Knysh observes that Sufism "has been inextricably entwined with the overall development of Islamic devotional practices, theological ideas, aesthetics, and religious and social institutions," making the study of Sufism as an isolated phenomenon rather distorting.[61] It follows that the study of the artistic life in premodern Islam without considering Sufism is equally distorting. Yet public perception of Sufism in the Islamic world has dramatically changed in modern times, and the current apathy and suspicion are not reflective of its influential social role in premodern times. Once a dominant feature of Muslim society with eminent and influential figures, Sufism is today a dubious phenomenon in many Islamic countries. Ironically, Western modernity, which was behind the sociopolitical reform that led to the rejection of Sufism as a perpetuation of medieval superstitions, backwardness, and ignorance, has itself provided a new home for its resurgence. In recent decades Sufism has gained noticeable popularity in the West and been the subject of a growing scholarly interest. Most illuminating studies are currently being published in the West, and a large body of literature is already available in English and other European languages on its history, doctrines, terminology, techniques, and practices. Due to the complex nature of their work, however, many key Sufi texts have remained unavailable in good editions, let alone good translations, and in this study I am using a number of texts that are still in manuscript form.

Existing studies on Sufism are numerous, adequately covering its history, doctrines, and practices in the eastern, middle, and western parts of the Islamic world. Alongside the writings of the perennialists, there are the works of many outstanding scholars, such as Massignon, Corbin, Izutsu, Nicholson, Arberry, Dermenghem, Gardet, Rice, Meier, Filipanni-Ronconi, Chodkiewicz, Addas, Chittick, and Knysh.[62] Through their insightful and penetrating analyses, their interpretations and translations of original Sufi literature, these scholars have made enormous contributions to the scholarly coverage of the phenomenon—its personalities, techniques, terminology, doctrines, and historical development. Thus there is no need to rehash what many distinguished scholars have already covered eloquently and comprehensively. A snapshot of what Sufism is all about, however, would help the reader who might not be familiar with the literature.

Intimacy

Two citations separated by over a millennium illustrate the consistency of the ultimate aim of the Sufi path. The first is a holy tradition *(ḥadīth qudsī)*[63] by the prophet Muḥammad, while the second is a soliloquy by the eminent Sufi master ʿAbd al-Ghanī al-Nābulusī (d. 1731).[64] Both describe what Sufism is all about, one from God's point of view, the other from the Sufis' perspective. Sufis regard the message of this holy tradition as constituting the ultimate aim of the spiritual experience and the cornerstone of mystical knowledge. Speaking through the Prophet, God says: "Nothing is more pleasing to me, as a means for my servant to draw near unto me, than worship which I have made binding upon him; and my servant ceases not to draw near unto me with added devotions of his free will until I love him; and when I love him I am the hearing wherewith he hears and the sight wherewith he sees and the hand whereby he grasps and the foot whereon he walks."[65] It is this intimacy with divinity that underpins the Sufis' vision, aspirations, and devotional practices. In the following soliloquy, the late Damascene Sufi master al-Nābulusī, an ardent follower of Ibn ʿArabī, shows how the prophetic tradition is personalized in the mystical experience. Later on, I will show how such soliloquial engagement assumes geometric personality in the writings of ʿAbd al-Karīm al-Jīlī. Al-Nābulusī writes:

> My Lord said to me: "you are good for me." I said: "how can I be good for you while I am perishable?" He said: "nothing is good for me except the perishable." I said: "how can I be good for you when my character is bad?" He said: "I complement it with my good Character." He then said to me: "O my servant, I am you, but you are not me; O my servant, I am who exists, not you; O my servant, all people are the servants of my benefaction *(niʿma)*, whereas you are the servant of my Self *(dhāt)*." I said: "but how am I the servant of your Self?" He said: "you are the servant of Being *(al-wujūd)*, not the servant of a being *(al-mawjūd)*. Being is me, while a being is other than me,

because beings are by me and I am by my Self. This is the reason why I said to you I am Being." He then said to me: "O my servant, do not fear other than me, because I am that other. I am your Lord, appearing unto you by my subsistence in you. There is no divinity if it be not me, and no one is adored other than me. In any state, if by me that I make you rich, truly, then, I have made you rich; but if I do so by other than me, I then have made you poor; there is no divinity if it be not me." I said to him: "O Lord, how am I in your regard?" He said: "you are to me amongst those who are drawn near *(al-muqarrabīn)*, and so is everyone who loves you: I love you and love everyone that loves you." I said to him: "O Lord, what is the sign of your loving me?" He said: "it is my guiding you to what I love and am pleased with." I said to him: "O Lord, people are harming me." He said to me: "all of this is of benefit to you; look within yourself for the result of their harm, you become nearer to me, and inevitably you will rise above them."[66]

Ibn ʿArabī

Muḥyī al-Dīn Ibn al-ʿArabī (abridged as Ibn ʿArabī), the central figure of this study, is one of the most prolific and influential figures in the history of Islam.[67] He was born in Murcia, Spain, in 1165 and moved as a child with his family to Seville where he received his education. He chose the Sufi path at a young age, and when he began to show signs of exceptional spiritual aptitude, his father set him up to meet a notable friend, the great peripatetic philosopher of Cordoba Ibn Rushd (Averroes, d. 1198), who had expressed interest in meeting the inspired young man. As described by Ibn ʿArabī, who apparently was aware of the set-up, the meeting was brief with only a few words being exchanged between the great Aristotelian master and the young beardless Sufi. Yet this brief encounter was to mark a historical moment that defined the distinction between philosophical speculation and mystical revelation in the attainment of truth.[68] This young man who dared to challenge the rational method of the great philosopher grew up to be no ordinary man. He became one of the greatest mystics of all times, who was known among the Sufis and other religious authorities as *al-shaykh al-akbar* (the "Greatest Master"). Like most Sufi masters, Ibn ʿArabī spent his life traveling throughout the Islamic world and making contacts with most prominent spiritual authorities of the time. On his way to the East, he went through North Africa, and together with a small group of followers he then traveled to the Ḥijāz, Palestine, Syria, Iraq, and Anatolia. He finally settled in Damascus where he died at the age of seventy-eight in 1240 and was buried in al-Ṣāliḥiyya at the foothill of Qāsiyūn in the suburb of Shaykh Muḥyī al-Dīn that carries his name to this day.

Ibn ʿArabī left us a large body of knowledge, which comprised 289 books and treatises, according to Ibn ʿArabī himself, four hundred and five hundred

Photo 1.1 The complex of Shaykh Muḥyī al-Dīn at the foothill of Qāsiyūn in Damascus.

according to al-Shaʿrānī and Jāmī.[69] Considering the complex nature of their topics and the many voluminous works he produced, such as *al-Futūḥāt al-Makkiyya*, the lost sixty-four- or ninety-five-volume interpretation of the Quran[70] and the lost three-hundred-chapter book *Manāhij al-Irtiqā'*,[71] even the most conservative number is extraordinary. In addition, one only needs to consider the numerous versions of, and commentaries on, his works that exist in various libraries, as well as the wide circles of followers he had attracted to appreciate the significance of his thought and teachings both for Islam in general and for Sufism in particular. Speaking of his standing in the Islamic world, Chittick says that "probably no one has exercised deeper and more pervasive influence over the intellectual life of the community during the past seven hundred years."[72] And in the introduction to his translation of *Fuṣūṣ al-Ḥikam*, Austin regards Ibn ʿArabī as representing "a culmination not only of Sufi exposition but also, in a very significant way, of Islamic intellectual expression."[73] Ibn ʿArabī's posthumous image

Photo 1.2 The dome over the tomb of Ibn ʿArabī in Damascus.

in the memories of subsequent generations, however, was not always one of veneration and esteem but was also shrouded with polemics and heresy. Al-Sakhāwī (d. 1497) compiled over three hundred legal opinions *(fatwa)* from various clerics issued between 1223 and 1490 on the status of Ibn ʿArabī.[74] In a recent study, Knysh has examined the perception of Ibn ʿArabī's personality and teachings by Muslim scholars in the four centuries following his death, showing that the wide fame he enjoyed was coupled with an aura of controversy.[75] In his opening remarks Knysh pointed out that the vast body of polemical literature associated with Ibn ʿArabī attests to his "abiding importance" in the Muslim world. This is not surprising, he says, because "from the 7th A.H./13th C.E. centuries onward practically every Muslim thinker of note took it upon himself to define his position vis-à-vis the controversial Sufi master."[76]

Through his works Ibn ʿArabī presents a complex cosmology with an intricate ontological structure. His ontology brings together a wealth of philosophical, theological, scientific, linguistic, metaphysical, and mystical knowledge, weaving them together into a cohesive, multidimensional whole with a unique level of intricacy and profundity. Interest in Ibn ʿArabī's writings and teachings has increased in the last few decades particularly in the West. The difficult and complex nature of his texts, however, has kept many of his works and particularly *al-Futūḥāt* largely inaccessible to the non-Arabic reader. Yet, several profound studies of Ibn ʿArabī's philosophy and mystical knowledge are available in European languages.[77] In two major books, Chittick has undertaken the daunting task of providing extensive translations of Ibn ʿArabī's magnum opus, *al-Futūḥāt.* His commentaries offer valuable insights into the difficulties involved in interpreting and translating such complex works. In dealing with original Sufi texts, I have benefited from the indispensable insights and excellent translations available in Western literature.

While dealing with a range of Sufi texts, the works of Ibn ʿArabī—by virtue of their richness, comprehensiveness, and coherence—are taken to represent the Sufi worldview in its utmost maturity and complexity. In this I run the risk of obliterating differences and variations in Sufi thoughts and teachings and of overstating Ibn ʿArabī's centrality and influence in the Islamic tradition. I acknowledge, however, that Ibn ʿArabī's ideas did not emerge in a vacuum and that he absorbed and represented much of what earlier and contemporary Sufis had established. His pivotal position comes from the inspirational power, comprehensiveness, cogency, and profundity of his syntheses, which, as many studies have shown, were instrumental in perpetuating and universalizing Sufi ideas in a coherent way. Also, Sufi conceptions of the genesis, structure, and layout of the universe reveal a remarkably consistent core, with variations being traceable mainly in modes of expressions. Ibn ʿArabī's monumental work, *al-Futūḥāt*, the main focus of this study, is known to have served as a main reference on Sufi ontology and cosmology for subsequent generations.[78]

Symbolism: A Sufi Perspective

A marginal commentary, attributed to Saʿd al-Dīn al-Ḥamawī (d. 1252/3), on a manuscript copy of Ibn ʿArabī's *Inshāʾ al-Dawāʾir* compares the entire world to the eye of God that never sleeps. The upper eyelid is compared to the upper world, the lower eyelid to the lower world, the eyelashes of the upper eyelid to the angels that dwell in the upper world, the eyelashes of the lower eyelid to the humans that dwell in the lower world, the iris to the Universal Soul, the white to the Universal Spirit, and the light whereby the eye sees to God.[79] This poetic imagery depicts eloquently the hierarchical structure of the world as conceived in premodern Islam. The notion of symbolism hinges on this hierarchical conception in constructing its ontological links between the lower and the higher, the sensible and the intelligible.

Premodern Islamic cosmology depicts a multilayered picture of the universe with each layer having its own inhabitants and objects. The earth as a central layer is covered by seven hemispherical heavens resting on seven infraterrestrial earths. This conception might have derived from a pre-Islamic mythology, yet it owes its continuity within the Islamic tradition to the Quranic references and prophetic traditions.[80] There are numerous descriptions reported in various forms after the Prophet or his immediate companions that elaborate this conception to considerable detail. Some will be discussed later.

The multilayered world depicted in premodern Islamic cosmology is inhabited not only by humans but also by other intelligent creatures: the angels and the jinn. Several traditions describe each layer in the celestial and infraterrestrial world to be an exact replica of the earth with all its existents. In a tradition describing the heavenly prototypes of the Kaʿba, the celebrated Prophet's companion Ibn ʿAbbās is reported to have described the existence of fourteen Kaʿbas, in addition to the known one, seven above and seven below, located exactly one above the other so that if they collapse they will fall one on the top of the other. He is also reported to have said that in each layer there are also creatures just as those on earth and that there are even other Ibn ʿAbbās just like him.[81] Alongside this one finds numerous narratives and traditions that describe the various ways of communication between the inhabitants of these worlds, revealing a preoccupation with the nature of the other "inhabitants" and their relationships to corporeal beings. In their treatise on the formation of animals and their kinds, Ikhwān al-Ṣafā' writes:

> Then know, O just king, that these forms, shapes, structures, and attributes, which you see in the world of bodies and material substances, are symbols, similitudes, and colors of those forms that are in the world of spirits, save that the latter are luminous and subtle, while the former are dark and dense. The relationship of the former to the latter is as the relationship between the paintings on the surfaces of boards and walls to the actual forms and shapes of the animals with flesh, blood, bones, and skin. Those forms in the spiritual world are the movers while these are the moved. As for the ones below, they are motionless, speechless, sensible, decayable, corruptible, and perishable, while those [above] are rational, intelligible, spiritual, invisible, and durable.[82]

This reminds us of Plato's cave parable wherein humans in their corporeal experience are portrayed as being twice removed from the real forms. In their daily experiences, Plato tells us, people deal only with the *sensible* shadows of the figures, which represent the *real*, the *intelligible* forms that reside outside the restraining boundaries of the cave of physical existence. The Platonic-Aristotelian ontology, which Muslim philosophers inherited, distinguishes between the sensible and the intelligible. This philosophic distinction was legitimized by the Quranic distinction between the *seen* and the *unseen*, within the framework of which symbolism became a necessary means of communication between two distant, yet ontologically related, domains of being.

Definitions

It may be easy to recognize in contemporary Arabic architectural discourse the equivalents of the terms *symbol* and *symbolism*—*ramz* and *ramziyya*—but once we shift our focus to premodern literature the task becomes much harder. This is because neither term was used consistently in the same modern sense, nor was either specifically associated with architecture. In fact many terms were used to denote the meanings of *sign* and *symbol*, without a sharp semantic distinction between the two. The Arabic terms *āya, ramz, ishāra, ʿibāra, mithāl*, and *dalīl* are all used to denote various shades of both 'sign' and 'symbol.' All are used in the Quran; however, the term *āya* is the most nuanced and frequently used. *Āya* literally means "mark" or "sign" but is most commonly used to refer to a Quranic "verse." Every Quranic verse is called "*āya*," Ibn Manẓūr explains, because "it is like a sign by which one is led to another as the road's flags that are erected for guidance"; also because "it is a group of letters of the Quran"; and because "it is a sign of discontinuity between two successive speeches."[83] These definitions suggest that *āya,* as a "verse," refers to a distinct group of letters and words that convey a particular meaning that is conducive to a following one and so on in a sequential manner, leading to the final meaning, in the same way road's flags guide people to their destination. The most potent meaning of the term, however, is "symbol" as in the widely quoted verse: "We shall show them our *symbols (āyāt)* on the horizons and within themselves until it will be manifest unto them that it is the Truth" (41:53). In presenting all created things as symbols, this verse can be seen as the cornerstone of the Islamic notion of symbolism. *Āya* also means "wonder," hence God's wonders are his *āyāt*. *Āya* is also used in the sense of *ʿibra*, meaning "example," "warning," "lesson," and "reminder": "Verily in Joseph and his brethren are *examples (āyāt)* for the inquiring" (12:7).

The terms *ramz* and *ishāra* denote other shades of the meaning. Literally, both words mean "a gesture made by the hands, eyes, eyebrows, lips or mouth with the intention of conveying something that could otherwise be expressed verbally."[84] *Ishāra,* a frequently used term for 'symbol,' has in addition two interesting derivatives, *shāra* and *shawra*, connoting the idea of beauty. Speaking of the prophet Zakariyyā, the Quran says: "you shall not speak unto mankind . . . except by *signs (ramzan)*" (3:41). Symbols, Ibn ʿArabī explains, are "dwellings" that enable us to reflect upon such things as divine unity, first Intellect, divine Throne, the science of representation *(ʿilm al-tamaththul)*, God's wonders *(āyāt)*, and so on.[85] Through these "dwellings," he says, the reflective minds find references or proofs (*dalāʾil*, sing. *dalīl*) for a multitude of interpretations.[86] In this sense, *ramz* stands for "symbol" in that it points to another context that is not apparent at the level of immediate experience. Ibn ʿArabī elaborates further, drawing our attention to the parallel between *ramz*, "symbol," and *lughz*, "riddle," in order to show the double function of symbols: guiding and misguiding, revealing and concealing. This is especially the case with *lughz*, which is a form of speech the outward meaning of which conveys a sense not intended by the speaker, making it prone to misunderstanding. This is the condition of the world, Ibn ʿArabī says. God founded the world for people to seek him, but they instead became preoccupied with the world itself, so they misunderstood the *lughz* of the creation and defied the intention of the founder.[87]

Following the etymological meaning of 'symbol,' from Greek *sym+ballo* "to throw together," "suggesting the way in which the symbol carries the mind to its referent,"[88] the term *ʿibāra* conveys a closer meaning. *ʿIbāra,* "expression," comes from the verb *ʿabara*, "to cross" and "to interpret," from which comes the word *ʿibra*, "lesson" or "wonder": "God causes the revolution of the day and the night. Herein indeed is a *lesson* (*ʿibra)* for those who see" (24:44). The trilateral root *ʿ.b.r.* means literally "to cross from one side of a river, valley, or road to another." A common meaning of *ʿabara* is "to interpret," "to expound," particularly dreams and visions, as in "expound for me my vision, if you can *interpret* dreams" (12:43). The expounder is *ʿābir*, "one who crosses," since in expounding one crosses from the outward to the inward side of the subject in order to reveal its hidden meaning. Ibn ʿArabī says that to every sensible form God has attached a spiritual meaning toward which one should cross by interpretation.[89] The semantic connection between "crossing" and "interpreting" is clearly expressed in a tradition that says: "O God, render us amongst those who interpret and understand *(yaʿbar)* the world and not amongst those who merely cross it *(yaʿbur)*."[90]

Plato features prominently in Islamic philosophy, and his doctrine of forms was the subject of thorough discussion by various philosophers.[91] The terms that were appropriated for the word "form" were *ṣūra* and *mithāl* or *mathal*, with the latter being more frequently used. The word *mathal* means "likeness" and "similitude," from *mithl,* "look alike." It is extensively used in the Quran in the sense of symbol: "Such *similitudes (amthāl)* we coin for mankind haply they may reflect" (59:21). *ʿĀlam al-mithāl*, the "world of similitudes" or the "realm of images," is a product of medieval Muslim mysticism similar, in many respects, to Plato's intermediary world. *Mithāl* also means "matrix" or "mould," according to which a design is made, and the derivative *timthāl* means a "statue" or an "image" made to resemble a creature. The *timthāl* of a thing also means its "shadow." *Mithāl* equates 'symbol' in the sense of being a shadow of a higher reality revealed in a sensible form. Yet *mithl*, "likeness," is also used in a different sense. In a tradition the Prophet is reported to have said: "Surely, I have been given the book and its likeness *(mithl)* with it." This is interpreted as being given the Quran along with the power to expound its inward meanings.[92]

Finally, the concealed meaning or significance of a symbol is often referred to as *sirr*, "secret" and "mystery," thus pointing to the intellectual effort required for the discovery of what is not immediately available. This brief survey of Arabic terms shows the range of meanings and concepts one encounters in premodern Islamic literature when dealing with the notion of symbolism. In the following I will locate these terms in a wider conceptual context.

The Seen and the Unseen

Religious worldviews hinge on an axiomatic premise that the world is made up of physical and spiritual realities, of visible and invisible entities. This premise underlies the fundamental beliefs in God, prophets, and holy scriptures that presuppose a kind of unseen, supranatural presence. In proclaiming itself to be the primordial as well as the last religion *(al-dīn al-ḥanīf),* Islam shares with other religions this view of the world. Both the Quran and the prophetic traditions speak of the "unseen" and "seen" worlds *(al-ghayb wa al-shahāda)*. The Quran stresses this polarity, describing God as "the Knower of the unseen and the seen" (13:9) and to him "belongs the unseen of the heavens and the earth" (16:77). The Quran repeatedly reminds the Muslims that no one knows the unseen except God: "And with him are the keys of the unseen. None but he knows them" (6:59). Yet aspects of the unseen can be revealed: "This is of the tidings of the unseen *(ghayb)*. We reveal it unto you" (3:44). The Quran demands that Muslims believe in the unseen and strive to gain knowledge of it by means of the seen: "This is the Scripture whereof there is no doubt, a guidance unto those who ward off evil. Who believe in the unseen" (2:2–3).

The Quran presents the seen as the world of the outward *(ẓāhir)* that is readily accessible to everyone. Therefore, no privilege or reward is promised in believing and participating in it. The unseen, by contrast, is the realm of faith that endows the believers with privileges and renders them worthy of reward: "Gardens of Eden, which the Beneficent has promised to his worshipers in the unseen. His promise is ever sure of fulfillment" (19:61). Ephemeral, transient, and perishable, the seen derives meaning and subsistence from the unseen, and its real value lies in being the necessary pathway to the unseen. The Quran likens this dependency to a plant flourishing in the rain but dying as soon as its source of life ceases to fall (10:25).

The seen is the world of natural realities that can be known directly through sense perception, whereas the unseen is the world of spiritual realities that can only be grasped by imagination. To help human imagination gain insight into the unseen, religious teachings have resorted to analogy and metaphor. The efficacy of analogy, as an illustrative and cognitive tool, hinges on the ontological link between the embodied and the abstract. By means of analogies human imagination is given access to the abstract through the mediation of the embodied. Analogy is thus the cornerstone of religious expressions that are concerned with spiritual phenomena. The Quran uses many tangible examples from the seen to explain or describe matters of the unseen:

> If all trees in the earth were pens, and if the sea eked out by seven seas more were ink, the words of God could not be written out to the end. (31:27)
> Do you not see how God cites a symbol: a good word as a good tree, its root set firm and its branches in heaven. (14:24)

In the first example the incomprehensible infinity of God's words is brought closer to human understanding by using the analogy of trees and seas as pens and ink. In the second, the verse relates "a good word" to "a good tree," so that we may understand the nature of the divine word by means of the given description of the tree. To say "a good word as a good tree" is to transfer the known information associated with the tree to the unknown concept of the 'word.' It is to try to understand the concept of the word by means of the concept of the tree. This mental activity involves identifying some underlying structural similarities between the concept of the word and the concept of the tree, and of transferring them from one onto the other. Religious understanding of spiritual realities hinges on the efficacy of such analogies, and symbolic reasoning relies on and promotes similar modes of thinking. In constructing ties between the divine and human modes of existence, analogical reasoning operates in the paradoxical space that lies in between the contrasting dimensions of analogy: *tashbīh* and *tanzīh,* "likeness" and "transcendence."[93]

In the often quoted Quranic verse: "We shall show them our symbols on the horizons and within themselves until it will be manifest unto them that it is the Truth" (41: 53), Ibn ʿArabī says, God alludes to his symbolic presences in all created things.[94] These symbols are available to humans in their daily experience of sensible things, be they "within themselves" or "on the horizons," that is, in the outside world. Their function is to give clues to direct the mind toward that which lies beyond the immediate attractions of the sensible and the visible, for "God coins the similitudes for mankind in order that they may reflect" (14:25). The Quran speaks of different kinds of analogies and symbols for different kinds of people: the contemplative, the faithful, the intellectual, and so on.[95]

> And of his symbols is this: he created you of dust, and behold you human beings, ranging widely!
> And of his symbols is this: he created for you helpmeets from yourselves that you might find rest in them, and he ordained between you love and mercy. Herein indeed are symbols for folk who reflect.
> And of his symbols is the creation of the heavens and the earth, and the difference of your languages and colors. Herein indeed are symbols for men of knowledge.

> And of his symbols is your slumber by night and by day, and your seeking of his bounty. Herein indeed are symbols for folk who heed.
> And of his symbols is this: he shows you the lightning for a fear and for a hope, and sends down water from the sky, and thereby quickens the earth after her death. Herein indeed are symbols for folk who understand.
> And of his symbols is this: the heavens and the earth stand fast by his command, and afterward, when he calls you, surely, from the earth you will emerge. (30:20–25)

> And of his symbols are the night and the day and the sun and the moon. Adore not the sun nor the moon; but adore God who created them, if it is in truth him whom you worship . . .
> And of his symbols that you see the earth lowly, but when we send down water thereon it thrills and grows. (41:37/39)

> And of his symbols is the creation of the heaven and the earth, and of whatever beasts he has dispersed therein . . .
> And of his symbols are the ships, like banners on the sea; if he wills he calms the wind so that they keep still upon its surface. Herein indeed are symbols for every steadfast grateful (heart). (42:29/32–33)

These Quranic references identify the "alphabet" of a special language whose existence is indispensable for conveying the divine message to humans. The necessity of this special "language" emerges from the incompatibility in nature between the communicants as well as the richness and intensity of the divine revelations, which cannot be adequately expressed in an ordinary language.

Distance and Deficiency

In *Maḥāsin al-Majālis* Ibn al-ʿArīf (d. 1141), the celebrated Andalusian Sufi master, cites an intriguing aphorism on symbolism. Using the term *ishāra*, he writes that "a symbol is a call from a distance and a disclosure of an essential deficiency."[96] This seemingly obscure statement sums up eloquently the Sufi view of symbolism: at once an intimate *call* and a *disclosure* of one's limitations. In the *Futūḥāt* Ibn ʿArabī clarifies this statement, explaining that as a means of communication the language of symbolism becomes a necessity under two conditions. One is when the communicants are so far apart that the voice of the speaker cannot reach the ears of the listener, although they can still see each other. In this case the only way for the speaker to convey their message is through displaying signs and symbols that the listener can understand. This is, he says, what the Sufis refer to by "a call from a distance." The other condition is when the communicants are close to each other but because of a deficiency in the listener, such as deafness, for example, they cannot hear the speaker's voice. Here again the language of symbolism becomes the only way to convey the speaker's message. It is with reference to this condition that the Sufis define symbols as "a disclosure of an essential deficiency." An example of this situation, Ibn ʿArabī explains, can be seen in the Islamic prayer, a condition wherein Muslims attempt to draw near to, and converse with, God, since Islamic prayer has been described in a prophetic tradition as a dialogue with God. Here human "deafness," as an inherent deficiency, is exposed when God replies to the praise addressed to him in their own tongue, saying: "God hears him who praises him" *(samiʾa -llāhu liman ḥamadah)*.[97]

Distance and *deficiency* are intrinsic to human nature when compared with that of the divine. The ontological difference between the creator and the creature is what manifests the polarizing distance that separates God and humans. God transcends human deficiencies and limitations, and it is this transcendence that makes the language of symbolism a necessity, not in respect of God being self-sufficient, of course, but in respect of him being the creator. In these ontological conditions human expression *(ʿibāra)* can only accommodate a limited spectrum of the divine truths, leaving much of the revealed knowledge beyond the linguistic grasp and hence uncommunicable directly through language. The ontological distance and deficiency disable humans from direct communication, calling for other, more effective means. The language of symbolism, by contrast, is less constrained than an ordinary language, and its vocabulary is more apt for communicating transcendental truths.[98] The efficacy of symbolism derives from the symbol's capacity to translate divine situations into human terms and vice versa, thereby bridging the gap created by distance and deficiency. Participating in both the divine and the human realms, symbols establish the necessary continuity between the order of the divine presence and that of human existence.

By means of symbols, communication with the divine can be ensured despite man's impeding ontological conditions, but so long as the communicants are in each other's range of "sight." Distanced and deficient, humans still have to "see" in order to understand. The act of "seeing" alludes here to the *in-sight* required to understand the meanings of signs and symbols. Distant from the symbol maker though they may be, those with in-sight are able to consume the distance, to draw near, to see and comprehend the intentions. Through symbols they comprehend what words cannot express or voice convey. In this sense, the significance of symbolism lies not in the symbol itself but in the meanings it communicates, the reality it unveils. Symbols are, therefore, not sought for themselves but for what they symbolize, for the insights they instil, the possibilities they disclose, and the meanings they deliver.[99] In a hierarchically ordered universe, the *unseen*, while setting itself apart from the *seen* by ontological distance and deficiency, projects a universal medium with an immense revelatory power, the medium of symbolism.

Shadows of the Immutable

In the *Futūḥāt* Ibn ʿArabī draws our attention to an intriguing phenomenon. As we all know, an object standing before a source of light projects a shadow that maps the object and remains attached to it. With the inclination of the light source, however, the shadow may extend well beyond the height of the object.[100] In fact, it can extend indefinitely despite being a projection of a definite height. Ibn ʿArabī prompts us to ask how the definite, the measurable, can project the indefinite, the immeasurable, and what does this allude to in the language of symbolism? These questions lead to interesting interpretations. The phenomenon of shadow has always intrigued the human mind. The shadow's curious relationship to the object it maps has often formed the object of philosophical reflections. In the famous cave narrative, Plato constructs his hierarchical structure of the universe using the shadow metaphor. The Quran also presents some thought-provoking statements on shadow: "Have you not seen how your Lord has spread the shadow *(ẓill)*? And had he willed, he could have made it still" (25:45).

ʿAbd al-Razzāq al-Qāshānī (d. 1330), a celebrated commentator on Ibn ʿArabī's works, reflects on the shadow phenomenon from the standpoint of universal manifestation. For a shadow to appear, he says, three things must necessarily exist: an object to cast a shadow, a ground whereupon the shadow to fall, and a light source to project the shadow. If we are to think of the creative process in terms of shadow projection, then the "object" can be taken to represent absolute Being, the "ground" on which the shadow falls to represent the archetypal essences of all possible beings, and the "light" that projects the shadow to represent the divine outward presence.[101] In the same vein of thinking, Ibn ʿArabī views the world as the exact shadow *(ẓill)* of the Absolute, manifesting at three different levels. Before the transcendental "light" of the Absolute extends the highest level of shadow, the archetypal forms, *aʿyān thābita*, "immutable essences" or "immutable entities."[102] From these essences extends the second level of shadow, the natural beings that project the immutable essences in embodied forms. From these embodied forms extends the third level of shadow, the sensible shadows that project the silhouette of natural bodies on sensible surfaces.

Theoretically coherent, this structure of reality raises some philosophical questions. First, one would want to know where the "ground" upon which the shadow falls has come from, and what is its ontological reality? Without a "ground," as al-Qāshānī observes, the shadow would remain potentially contained in the object; it would remain an intelligible occurrence as a tree within a seed. At the first and second levels, the Sufis argue, there is no distinction between the

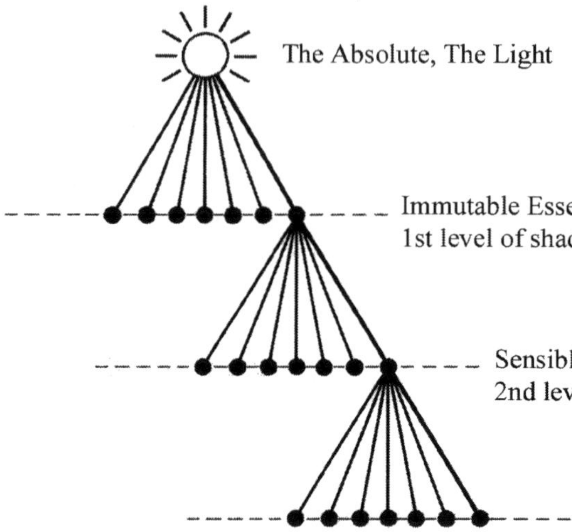

Fig. 1.1 The hierarchy of shadows according to Ibn ʿArabī.

shadow and the ground. Both the immutable essences and the concrete beings are at once the ground and the shadow.[103] Only at the third level the distinction occurs.

Insofar that the second phase of shadows is actualized in the tangible forms of all existents, Ibn ʿArabī argues, sensible existents can be regarded as the real shadows. The immutable essences, by contrast, are nonexistent sensibly: "they have not even smelt the fragrance of existence."[104] If we are to imagine the process of shadow projection as being instantaneous, then sensible beings can be seen as the ground on which the shadow of immutable essences falls. Twice removed from the source of light, they project the embodiment of the immutable essences into the physical world. This relationship is compared to that between, for example, the idea of wood-ness and its embodiment in every piece of wood, and in every chair, table or box made out of wood; or to the idea of quadrature and its embodiment in every square and in every paper, book, or house made in a square form. It is also like the quality of whiteness in every white color and of humanity in every human being; they are always the same, neither changing nor multiplying with the indefinite variety of white colors and humans that can possibly exist.[105] In all these cases none of the intelligible realities can be said to have formed part of the things in which they are manifested, nor, of course, can it be said that any of them has been multiplied by the multiplicity of forms in which it has appeared. The reality *(ḥaqīqa)* of wood-ness, quadrature, whiteness, and humanity is revealed equally in each embodied manifestation.[106]

Another set of philosophical questions can be raised from the Quranic perspective. If all existents in the heavens and the earth are no more than shadows, then how can one make sense of the Quranic verse that says: "And unto God prostrates whatsoever is in the heavens and whatsoever is in the earth of living creatures and the angels" (16:49); "and the stars and the trees prostrate (in worship)" (55:6). What is the ontological mode of "prostration" in this context? Ibn ʿArabī's answer is passivity. He interprets the act of prostration as the passive nature of shadow, always dependent on and determined by the reality it projects. This applies to all degrees of shadows. God refers to all existents as prostrating themselves before him because, as shadows, they project the original passivity of the immutable essences. When the shadows of the immutable essences come into existence in the form of heaven, earth, sun, moon, stars, mountains, trees, beasts, and every other being, he says, they naturally reflect the tendency of their archetypes.[107]

Through this ontological nexus between the lower and higher degrees of shadow Ibn ʿArabī explains how the sensible shadow of definite objects can be indefinite. In this he reveals how symbolism works by tracing back the process of universal manifestation. The shadow of natural bodies, in being an integral part of this universal hierarchy, acquires symbolic power. Although the last and farthest in the chain of differentiation, it nonetheless carries within it qualities that point toward its transcendental reality. Symbolism derives its effectiveness from this unbroken nexus of projections, and the symbol's efficacy lies in its capacity to bring into the phenomenal world a quality of the Absolute. The indefinite extension of shadow, Ibn ʿArabī explains, is a reminder of the infinite reality whence a being proceeded. Between two immeasurable shadows (that is, the immutable essence and the sensible shadow), we appear in measurable bodies, yet we remain bounded by two immeasurable presences. The extension of sensible shadow seeks that intermediary presence of the immutable essences, which in turn seeks the presence of absolute Being *(al-wujūd al-muṭlaq)*, which reveals itself through the attribute of "Light."[108]

The power of symbolism hinges on such ontological conception, wherein existential differentiations are projected as degrees of manifestation sequentially crystallized. Sensible beings, as symbols, are determined and conditioned by the intelligible realities they embody, revealing an ontological dependency on the higher degree of existence they reflect, just as the shadow is inherently dependent on the object it projects.[109] This ontological link that ties all levels of existence together into a continuous chain is reflected in the premodern Islamic view of making, wherein sensible objects take their final shape through an ontological sequence of differentiation, making it possible to trace a sensible form back to its original source.

Making and the Chain of Differentiation

Ibn ʿArabī's ontology of shadow, expressive though it may be, remains a broad-brush sketch of universal manifestation with only three ontological "rings" in the chain of differentiation. Around three centuries earlier, Ikhwān al-Ṣafāʾ provided a more detailed picture of the process of differentiation, which they discussed with reference to the notion of making at the human, natural, and divine levels. Relying on the premodern philosophical notions of form *(ṣūra)* and matter *(hayūlā)*, the Ikhwān established a nuanced chain of differentiation that ties human and natural artefacts up to the original source. Following the theoretical models inherited from the Greeks, most medieval Muslim philosophers worked with the notions of form and matter; however, the Ikhwān were probably among the earliest to articulate their pertinence to the arts *(ṣanāʾiʿ)* in a systematic manner. Their Pythagorean, Neoplatonist perspective is particularly instructive with regard to showing how the hierarchical conception of the universe influences the way in which the notion of making *(ṣanʿa)* is understood and theorized. They also show how the sensible-intelligible polarity plays out in the context of designing and how all fit in structured hierarchies of form, matter, and designed objects.

In their *Rasāʾil* the Ikhwān teach that every manifested object, whether artificial (man-made) or natural, necessarily comprises two fundamental components: form and matter. By "matter," the Ikhwān explain, philosophers mean generally "every substance *(jawhar)* that admits form"; and by "form" they mean "every shape and motif a substance is able to admit."[110] Since different things can be made from the same matter, they add, it is form that is considered responsible for the differences between things.

> Know that existents differ by form not matter. For we find many things that have one substance while their forms are different, such as a knife, a sword, an axe, a saw, and all that is made from iron, be it a machine, an instrument, or a container. The difference in the names of these things derives from their different forms and not from their different substances. With respect to their substance, the iron, they are all one. Likewise is a door, a chair, a bed, a ship, and all that is made from timber, the difference in their names derives from their different forms. As for their matter, the timber, they are all one. This is the way in which matter and form are considered to be in all artefacts *(maṣnūʿāt)*, since it is necessary for any artefact to have a matter and a form in its composition.[111]

The Ikhwān further explain that the process of making *(ṣanʿa)* involves necessarily two kinds of art: practical art *(ṣanʿa ʿamaliyya)* and theoretical art *(ṣanʿa ʿilmiyya)*.[112] The former refers to the sensible production of an artefact, whereas the latter refers to the knowledge that leads to such production, with both being subject to external forces that initiate and propel the operation of making. They make no distinction between making and designing for utilitarian and artistic purposes. In the epistle on practical arts the Ikhwān write: "Practical art is the artificer, the knower, externalizing the form that is in his mind and placing it in matter. The artefact is a whole made up of matter and form together. This begins by the influence of the Universal Soul on the process forced by the impetus of the Universal Intellect with the order of God—exalted be his praise."[113]

In this theoretical articulation the act of designing acquires cosmological dimensions, and the designed object acquires symbolic significance through the sequential process of differentiation that causes it to exist. Such conception renders all human or natural makings as stages of differentiation in a universal, existential process. In the original source everything is conceived as a non-differentiated totality, and the further bodies are from this source the more intricate and complex they become. The Ikhwān write:

> The shirt is a form with regard to the cloth, and the cloth is matter for the shirt; the cloth is a form with regard to the yarn, and the yarn is matter for the cloth; the yarn is a form with regard to the cotton, and the cotton is matter for the yarn; the cotton is a form with regard to the plant, and the plant is matter for the cotton; the plant is a form with regard to the *arkān* (elements), and the *arkān* are matter for the plant; the *arkān* are a form with regard to the [Absolute] Body, and the Body is matter for the *arkān*; the Body is a form with regard to the [Prime] Substance, and the Substance is matter for the Body. Likewise, the bread is a form with regard to the dough, and the dough is matter for the bread; the dough is a form with regard to the flour, and the flour is matter for the dough; the flour is a form with regard to the grain, and the grain is matter for the flour; the grain is a form with regard to the plant, and the plant is matter for the grain . . . This is the way in which form relates to matter and matter to form [in a sequential manner] until they terminate with the Prime Matter *(al-hayūlā al-ʾūlā)*, which is nothing but the form of existence that includes neither quality nor quantity. It is a simple Substance—without any kind of synthesis whatsoever—that is susceptible of all forms in a sequential order, as we showed, and not randomly. For instance, the cotton does not take on the form of the cloth until it has received the form of the yarn; and the yarn does not take on the form of the shirt until it has received the form of the cloth; likewise, the grain does not take on the form of the dough until it has received the form of the flour; and the flour does not take on the form of the bread until it has received the form of the dough. In this order matter takes on forms one after the other.[114]

As the binary principles of human making, form and matter are only a reflection at the human level of existence of a complex ontological structure and universal hierarchy. Making is viewed not to be an exclusively human activity: God and nature also make. Accordingly, the Ikhwān distinguish four kinds of making and four types of artefacts: human *(bashariyya)*, natural *(ṭabīʿiyya)*, psychic (*nafsāniyya*, referring to the Universal Soul), and divine *(ilāhiyya)*. This differentiation relates to the four different types of matter: matter of artificial work *(hayūlā al-ṣināʿa),* matter of natural work *(hayūlā al-ṭabīʿa),* Universal Matter *(hayūlā al-kull),* and Prime Matter *(al-hayūlā al-ʾūlā).*[115] The Ikhwān explain:

> The matter of artificial work is every body *(jism)* out of and in which an artificer works his art, such as the timber for carpenters, the iron for ironsmiths, earth and water for builders, the yarn for weavers, and the flour for bakers. Accordingly, it is necessary for every artificer to have a body to work his art from and in it. This body is the matter of artificial work . . . Natural matter is the four elements *(arkān)*. All that is found in the sublunary sphere, the animals, plants, and minerals, come from the elements and by corruption return to them. The active nature responsible for this process is one of the forces of the celestial Universal Soul . . . Universal Matter is the Absolute Body, from which is drawn the entire world, that is, the celestial spheres, the stars, the elements, and all beings. These are all bodies whose diversity derives from their diverse forms. As for Prime Matter, it is a simple, intelligible substance that cannot be sensed, for it is the form of being proper. It is the Original Identity *(al-huwiyya).*[116]

Made objects or artefacts correspond directly to this ontological structure, with each type of artefact corresponding to its respective level of matter. Human artefacts correspond to the matter of artificial work; natural artefacts to the matter of natural work; psychic artefacts to Universal Matter; and divine artefacts to Prime Matter. The Ikhwān further illustrate the four types of artefacts in the following examples:

> The human artefacts are those shapes, motifs, and paints, which craftsmen work in natural objects . . . The natural artefacts are the sensible forms of the animals, the diverse shapes of the plants, and the colors of the mineral's substances. The psychic artefacts are those such as the pattern *(niẓām)* of the four elements: fire, air, water, and earth, which are in the sublunary sphere, the composition of the [celestial] spheres, and the formal pattern of the entire world. The divine artefacts are the abstract forms without matters: the inventions by the inventor of all invented things . . . an existence from non-existence . . . a thing from no-thing; one impulse without time, nor place, nor matter, nor form, nor movement.[117]

Logically, the ontological differentiation of matter demands a corresponding differentiation of form. Such differentiation coincides with the process of, as it were, the "coagulation" or "condensation" of matter, that is, with the procession from Prime Matter, the simple spiritual form that possesses no qualifications, to the qualified and perceptible matter of particular objects.[118] The coagulation of matter is effected by the successive manifestation of three forms or ideas: Original Identity *(huwiyya)*, Quantity *(kammiyya)*, and Quality *(kayfiyya)*. The Ikhwān explain:

> When the Original Identity admits Quantity it becomes the Absolute Body that is referred to as having three dimensions—length, breadth, and depth. And when this Body admits Quality, that is, the shape, like circularity, triangularity, quadrature, and others, it becomes a special body referred to by the name of whatever shape it takes. Thus Quality is as number 3, Quantity as 2, and the Original Identity as 1. Just as 3 comes after 2, so does Quality come after Quantity; and just as 2 comes after 1, so does Quantity come after the Original Identity, which precedes Quantity and Quality as 1 precedes 2 and 3 and all the numbers.[119]

Al-huwiyya, interpreted here as "Original Identity," refers to the primordial substance that precedes all substantial differentiation. It identifies the intimate state of divine Being designated by Sufis as *huwa* (he), hence the name *huwiyya*. The Ikhwān explain: "Identity, Quantity, and Quality are simple, intelligible forms that cannot be sensed. When related to one another, some act as matter while others as form. Quality is a form in Quantity, while Quantity is matter for Quality. Quantity is a form in Identity, and Identity is matter for Quantity."[120] This is the archetypal model of the chain of differentiation that establishes an unbroken chain, linking human, natural, and divine acts of making.

Along their chain of differentiation, the Ikhwān further articulate the ontological nature of objects with regards to their forms. They distinguish two kinds of forms: necessary and complementary. The formation of any sensible object, they posit, necessarily comprises *ṣūra muqawwima,* "necessary" or "fundamental form," and *ṣūra mutammima,* "complementary" or "perfective form." These can also be referred to respectively as "*ṣūra jawhariyya*," "essential form," and "*ṣūra ʿaraḍiyya*," "accidental form."[121] Both the necessary and complementary forms can be understood in a universal as well as particular sense. In the universal sense, the necessary form is the foundational quality of all bodies in space, that is, of having the three dimensions of length, breadth, and depth.[122] The complementary form is what complements the necessary form, giving the object its sensible characteristics and bringing it to its most perfect state. It includes such things as configuration, movement, light, purity, and so on. Thus viewed, an object has one necessary form but many complementary ones. The Ikhwān explain: "The necessary form of a thing is that which brings with its parting the thing's matter the termination of its existence. Whereas the complementary form is that which brings the thing to the noblest states it is capable of reaching, and whose parting the matter of the thing brings no termination to its existence. For example, when the forms of stillness and movement part a body, its existence is not terminated; but when the form of the three dimensions of length, breadth, and depth parts the matter of a body, the body ceases to exist."[123]

In its particular sense, the Ikhwān's concept of 'necessary and complementary form' finds applications at the level of human making or art. One form can at the same time be necessary and complementary with regard to different aspects of the same artefact. For example, the form of a shirt, which manifests by the act of sewing, is a necessary form with regard to the essence of the shirt but complementary with regard to its material. For as long as the shirt remains in the form of unsewn pieces of material, the shirt as such does not exist, but the material itself does. And insofar as the action of sewing brings the material to a better state as a shirt (since any wrought object is viewed to be better than its raw material base), the form of the shirt is complementary with regard to its material. The same can be said about the form of the material that manifests by the act of weaving: it is a necessary form with regard to the essence of the material but a complementary form with regard to the yarn. For once the form of weaving parts the yarn there would be no material as such, but the yarn would remain.[124]

The Agency of Imagination

The Islamic concepts of 'form' and 'matter' directly relate to the concept of 'imagination' *(khayāl)*, which is often described as a rarefied substance capable of admitting and manipulating sensory forms once abstracted from their material entrapment.[125] Imagination is generally thought of as having mnemonic, representational, and creative functions; however, it is its ontological status that provides indispensable insights into its agency in the Sufi view of symbolism. Within the sensible-intelligible polarity Muslim philosophers and mystics articulated a concept of imagination with two distinct functions: dreaming and imagining.[126] Dreaming is an involuntary act that formed the focus of mystical and rational sciences concerned with visionary experiences, dream interpretation, and divine inspiration. Imagining is, by contrast, a voluntary, multifaceted function that formed the focus of a range of sciences concerned with the nature of the soul and its intellectual faculties. With regard to creativity, the act of imagining was viewed as involving the retaining of images perceived through the senses by memory *(al-quwwa al-ḥāfiẓa)*, the recalling of images when they are no longer in contact with the senses, and the composing of new images by the form-giving faculty *(al-quwwa al-muṣawwira)*. Imagination was also seen to mediate both analogical reasoning and symbolic representation by bringing abstract concepts and sensory forms together in meaningful relationships. In addition to being an essential cognitive instrument, Sufi viewed imagination as the creative cause of our existence and the powerful agency that enables us to remain in contact with the infinite and the Absolute. "He who does not know the status of imagination," Ibn ʿArabi asserts, "is completely devoid of knowledge."[127]

Imagination was viewed to operate within the ontology and hierarchies of form and matter. In the subtle matter of imagination, forms become free from the restraining forces of their sensible matter, and as such they can be easily merged and fused into one another. In this sense, the creative nature of human imagination becomes its capacity to deal with and manipulate the abstracted forms in whatever way it wishes. According to the Ikhwān, this is how, for example, "one can imagine a camel standing on the top of a palm tree or a palm tree on the back of a camel, a bird with four legs, a horse with two wings, a donkey with a human head."[128] In this sense, the creative power of human imagination can be explained with reference to its ability to deconstruct available forms and reconstruct new familiar and unfamiliar forms that embody new meanings.

Ibn ʿArabī further explains that although the senses might not have perceived the new images composed by the human imagination in their composed form, their elementary components must have been sensibly perceived. Human imagination, he stresses, cannot deal with anything that does not have, in part or in whole, sensible form.[129] This means that our individual imagination is inextricably bound to the sensory world. All the elementary data that fill our imaginary reservoir are extracted by our senses through contact with the phenomenal world so that our imagination has no power of *creatio ex nihilo.* We are, nevertheless, able to synthesize things creatively in our imagination according to new, unfamiliar patterns, as when, for example, we form an image of a creature half man and half horse even though there is no sensory model for such an image. Neither the image of man nor that of the horse, however, is a pure product of our imagination, but are both no more than imaginary reflections of sensory prototypes. Al-Ghazālī maintains a similar argument, showing the dominant role of vision over other senses in the process of imaginary constructions and interpretations.[130]

Ibn ʿArabī's teachings have given the concept of imagination ontological dimensions. Ibn ʿArabī differentiates three entities at the highest universal level: *al-wujūd al-muṭlaq*, "absolute Being," *al-ʿadam al-muṭlaq,* "absolute non-Being," and a *barzakh*, "mediator," that delimits the two. The first is the unrestricted existence of God, the necessary Self-existent; the second is the non-Self-existent; and the third is the intermediary domain of archetypes of all possible existents *(al-mumkināt).*[131] The intermediary world, or the "isthmus," as Ibn ʿArabī calls it, derives from the Quran, which makes more than one allusion to its nature: "He has loosed the two seas to

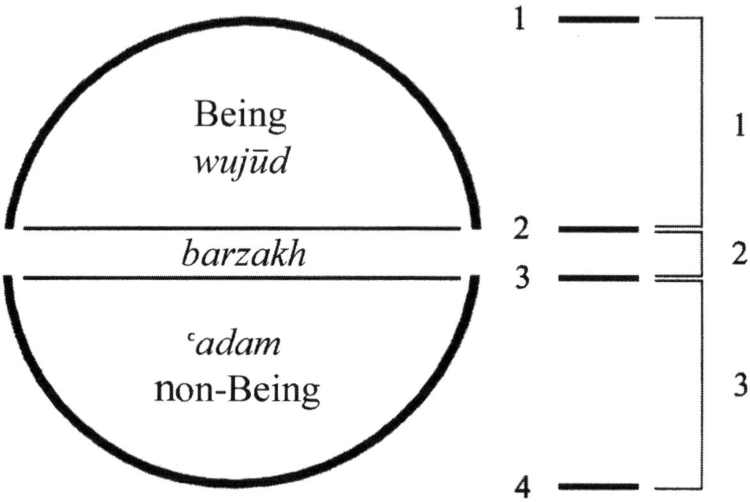

Fig. 1.2 Triplicity and quadrature underlying the order of universal manifestation.

meet, yet between them stands an *isthmus (barzakh)*, which they cannot overrun" (55:19–20); "It was he who brought forth the two seas; the one sweet and fresh, the other salt and bitter, and set between them an *isthmus (barzakh)* and an insurmountable barrier" (25:53). These verses portray an image of two integral domains, the "two seas," that are at once related and separated by an intermediary "barrier." Just as the borderline between light and shadow, the barrier through its unitive-separative nature brings together the two neighboring domains into a productive relationship. Acting as a common horizon that reflects the realities of both bordering worlds, the *barzakh* is the medium through which the delivery of the world from potentiality to act is effected. The world becomes, as it were, the "child" born from the fruitful marriage of absolute Being and absolute non-Being.[132]

Carrying the "genetic structure," so to speak, of its "parents," the created world is likewise tripartite. It is made of higher, lower, and intermediary realms. The higher world is the realm of the unseen, of spiritual being, of angelic forms, and of abstract meanings; whereas the lower world is the realm of the seen, of corporeal being, of the senses and sensible forms, and of natural bodies. In between there lies the third realm, which Ibn ʿArabī calls "*ʿālam al-khayāl*," the "world of imagination." It is the realm that combines the characteristics of its two bordering worlds; it is the place where the spirituality of the unseen is integrated into the corporeality of the seen to create the subtlety of the imaginary. It is the ontological level at which spirits manifest in sensible matrices and abstract meanings take on bodily forms.[133] As the dream world, the world of imagination is at once real and unreal, wherein things feel touchable yet remain unreachable. Imaginable forms, like dreams, have an apparitional or phantasmal quality: they are perceivable, meaningful forms yet without physical presence. They are neither purely sensible nor purely abstract. Like an image in a mirror, it is visible, yet not there; it is visible but without a body; and like an illusive mirage, it is there yet it can never be reached.[134] These are the qualities of the isthmian world of imagination. The elusive nature of imaginary forms derives from its intermediary function between the pure and the gross, the spiritual and the physical, the meaningful and the sensible. The world of imagination is the level of existence where this duality is resolved: where the pure is embodied and the body is purified. Imagination is the world where meaning and form marry, generating a new world that is at once uniting and separating its parental domains, just like the twilight zone, which unites and separates light and darkness.

"Know that you are an imagination," Ibn ʿArabī says, "and everything that you perceive, and of which you would say 'this is not me,' is also an imagination. So the whole being is an imagination within an imagination."[135] Here Ibn ʿArabī alludes to more than one level of imagination. In keeping with his universal hierarchy and concept of 'shadow,' Ibn ʿArabī distinguishes three ontologically different levels of imagination: a transcendental, unrestricted imagination, *khayāl muṭlaq*, "absolute imagination"; an all-encompassing imagination, *khayāl munfaṣil*, "detached imagination"; and an encompassed imagination, *khayāl muttaṣil*, "attached imagination."[136] The notion of imagination, however, still designates two different, yet related, things: a state of being and a creative capacity. As a creative capacity, the "detached" and "attached" polarity differentiates between the divine and human modes of creativity. Attached imagination, Ibn ʿArabī explains, designates our ability to abstract, conceive of, and manipulate forms. It is our imagining faculty operating within the human psychological framework. It is called "attached" because it is, as Corbin explains, "an imagination *conjoined* to the imagining subject and inseparable from him."[137] Corbin refers to it as "dependent imagination" because it depends in its existence on "detached imagination" which he translates as "autonomous." "Detached" or "autonomous imagination," by contrast, refers to a higher creative capacity that causes all imaginable forms to exist. The forms conceived by "attached imagination" are extracted by the senses from natural forms, which are a part of the cosmic forms that belong to a self-subsisting presence independent of the imagining subject.[138] Detached imagination is divine imagination, God imagining the world; it is the presence of the world in the divine mind. Attached imagination is human imagination, man imagining the forms of existents brought into existence by the creative power of divine imagination; it is the presence of things in the human mind.

Attached imagination depends upon detached imagination, and the human act of imagining is no more than a participation in the latter. Ibn ʿArabī says: "from this detached imagination the attached imagination derives."[139] While both have creative power, detached and attached imaginations are fundamentally different: one is permanent and the other transitory. The permanence of the detached derives from divine eternity, whereas the transience of the attached is a reflection of the humans' ephemerality. "The distinction between attached and detached imagination is that the attached vanishes when the one who is imagining vanishes; the detached is an essential presence *(ḥaḍra dhātiyya)* permanently receiving meanings and spirits so that it embodies them by its special capacity."[140]

In that human imagination has the power of participating in the world of detached imagination, it is capable of composing an infinite number of different kinds of images; but in that it is confined to the sensible domain, its elementary data are limited. It follows that out of the limited data that human imagination has at its disposal humans have the capacity to synthesize as many forms as there are possibilities latent in the world of detached imagination. Here one observes that compared with the divine imaginative power human capacity fades into insignificance; and despite the infinite possibilities available to God the forms composed in the divine imagination and revealed to us through the process of manifestation follow certain, specific, and well-defined patterns (the natural patterns that govern the created world). The questions may then be asked: If the human imagination is capable of composing an indefinite variety of forms, which includes, as the Ikhwān put it, "images that have corresponding realities and others that do not," what is the value of the composed forms that have no corresponding realities or archetypal patterns? What is the value of the imaginary form composed of, say, half man and half horse, if it is a mere possibility?[141] There are no simple answers to such questions, of course, for aesthetic as well as religious values play a role. Although Ibn ʿArabī does not address this question, some cues can be found in his teachings. Imagination, he teaches, is corrupt when it does not conform to the realities. An imagined form has to correspond to its reality to be regarded as falling in the category of "mental existence" *(al-wujūd al-dhihnī)*. "Mental existence," he writes, "is the known *(al-maʿlūm)* being imagined in the soul according to what it is in reality; but if the conceived image does not conform to reality, that would not be an existence of the known in the mind."[142]

As an all-encompassing, permanent presence, the world of detached imagination can then be seen as governing the human attached imagination by setting an immutable code for it. Such a code, whose content is made up of the cosmic realities, is considered necessary to prevent the human imagination from degenerating into a kind of fantasy. Participating in this realm of realities, human imagination can become either a valuable source of knowledge when it complies with the realities of that code or can become corrupt fantasy when it does not. Furthermore, an ignorant person, Ibn ʿArabī teaches, is one who speaks of, or believes in, what he forms in his soul while that which he has formed has no corresponding form other than itself. Any imaginary form that has no "existential presence" *(ḥaḍra wujūdiyya)* governing its existence is produced by ignorance, and the producer is ignorant. "And anything that has no form except in the soul of its speaker," he adds, "vanishes from existence with the vanishing of his saying or the vanishing of his memorizing what he may have fancied from his speech, for there is no existential presence that governs its existence."[143] With regard to its elementary components, however, whatever is composed in man's imagination necessarily has corresponding archetypes, for human imagination cannot escape the world of archetypes, which Ibn ʿArabī says contains the essences of all possible things. But with regard to a composed form, if it has no existing archetype, it would be insignificant, for it would point to nonexistence *(al-ʿadam)*. The sensible form produced according to an insignificant imaginary form has, according to Ibn ʿArabī, no "fatherly" principle determining its existence but only a "motherly" principle, the producer himself.[144]

Imagining, Knowing, and Making

In their definition of practical art, quoted above, the Ikhwān referred to the one who brings out the form from his mind and places it in matter as "*al-ṣāniʿ al-ʿālim*," "the artificer, the knower." They associate art *(ṣanʿa)* with knowledge *(ʿilm)*," assigning to knowledge as essential a role in the artistic operation as the skills of the artificer. Yet it is only natural for the Ikhwān to attribute knowledge to the artificer since in their view the forms retained in the artificer's mind are received through the act of knowing. Knowing, they say, is nothing but "the soul imagining the form of the known";[145] and "knowledge is nothing but the form of the known [retained] in the soul of the knower"; whereas "art *(ṣanʿa)* is nothing but the bringing out of this form, which is in the soul of the artificer, the knower, and placing it in matter."[146] Thus the artificer has necessarily to be a knower if he is to claim possession of any form in his mind. Such view makes art and knowledge an indissoluble whole. It also assigns to imagination an essential role in the human act of knowing, whereby the known becomes identical with the imagined forms of information imprinted in the knower's soul.[147] In *al-Risāla al-Laduniyya*, a treatise that deals with the nature of esoteric knowledge, al-Ghazālī gives similar definitions. "Know that knowledge is the imagining *(taṣawwur)* by the rational, tranquil soul of the realities of things and their abstract forms in themselves and in their qualities, quantities, substances and essences if they are simple. And the knower is the one who encompasses, conceives and imagines, and that which is known is the essence of the thing, the knowledge of which is engraved upon the soul."[148]

Both definitions conflate the acts of knowing and of imagining. Ibn ʿArabī, however, has a somewhat different view. For him, the act of knowing is clearly distinguishable from the act of imagining. Knowledge, he says, is not imagining the form or the meaning of the known because not every known can be imagined or every knower have the ability to imagine. If a knower can imagine the form of the known, this would point to his imaginative capacity, for there are things that can only be known without being imagined. The best example is God. Ibn ʿArabī writes:

> The conceiver and that which can be conceived—each is of two kinds. There is a conceiver that knows and has imaginative power and a conceiver that knows but has no imaginative power. There is also a conceived thing that has a form, according to which it can be known but not imagined by the one who has no imaginative power, and known and imagined by the one who has imaginative power, and a conceived thing that has no form and can be known only . . . Knowledge is not conceiving the form of the known, nor is it the meaning of the known being formalized. Because not every known admits form nor every knower is able to conceive of form. Form-conception is related to the knower through the latter's ability to imagine, and form is related to the known through the state in which it is accessible to imagination. And since there are knowable matters that are originally inaccessible to imagination, it is certain that they have no form.[149]

Thus, form, in relation to the known, is the state of the known that can be grasped and retained by human imagination. And since Ibn ʿArabī considers the data of our imaginary repertoire to be restricted to the confines of our sensory perception, human imagination is incapable of grasping anything that has no likeness, in whole or in parts, in the sensible world. It follows that the Quranic description of God as "Naught is as his likeness" (42:11) designates a reality to which human imagination cannot ascribe a form, for it has no likeness in comparison to which it may be grasped. This reality, however, can be known as it is in itself, that is, being without likeness.[150] Accordingly, knowledge for Ibn ʿArabī is "the acquisition, by the heart, of a particular matter according to what it is in itself, whether existent or nonexistent. Knowledge is the attribute that necessitates the act of acquisition from the heart, the knower is the heart, and that which is known is that acquired matter."[151]

Apart from the polemics on the nature and knowing and imagining, both views share some discursive grounds with regard to the notion of making. In Arabic "form" *(ṣūra)*, "imagining" *(taṣawwur)*, and the act of "drawing," "painting" and "forming" *(taṣwīr),* all derive from the verbal root *ṣawwara* which means "to form," "to configure," "to fashion," "to draw," and "to paint."[152] In the *Rasāʾil*, the Ikhwān deal with the notion of art, using the term *ṣanʿa* (plural, *ṣanāʾiʿ*), "craft" and "craftsmanship." The term derives from the verbal root *ṣanaʿa,* which has several related meanings: "to do," "to make," "to manufacture," hence, *ṣināʿa* is "industry," "to fabricate," and "to create"; hence, God is *ṣāniʿ*, the "artificer" of the world. In articulating their notion of art, they make no distinction between different kinds of making: everything that people make seems to have been viewed as a work of art. In their view there is no such thing as "fine arts" and "applied arts," nor is there any distinction between "artist" and "craftsman" or "art" and "craftsmanship."[153] Certain crafts, however, are considered to be nobler than others according to their subject matter. The craft of calligraphy, for example, deals with the sacred letters of the Quran and is, therefore, considered to be nobler than carpentry, which deals with timber.

The Ikhwān polarize art into *practical* and *intellectual* in accordance with the philosophical distinction between the practical and intellectual faculties of the soul. Ibn ʿArabī further articulates this view by polarizing forms into outward and inward. The outward forms are the tangible bodies together with all their sensible qualities, whereas the inward forms are the intelligible sciences, insights, and intentions that reside within the sensible forms.[154] Production occurs by the "marriage" of the practical and intellectual skills of the maker, whereby forms come into existence. Ibn ʿArabī views the intellectual skill as representing the active, determining, fatherly principle, whereas the practical skill as representing the passive, determined, motherly principle. An artefact then becomes as it were the "child" born from this fruitful relationship. This productive process applies at different levels in the artistic operation and could even be performed by two different people, if one person does not possess both the practical and intellectual skills, as, for example, is the case with a knowledgeable geometer and a skillful carpenter. Ibn ʿArabī explains:

> If a geometer *(muhandis)*, who is also a carpenter, is not skillful in practice, he may reveal the knowledge he has in the form of words to the hearing of one who is skillful in carpentry. This revelation causes a marriage relationship. The speech of the geometer is a father and the receptivity of the listener is a mother. The knowledge of the listener then becomes a father, and the organs of his body become a mother. And if you wish you may say that the geometer is a father, and the craftsman, who is the carpenter, in that he listens to what the geometer tells him, is a mother. Now if the geometer's speech causes an effect in the carpenter, then the geometer has imprinted that which he has within him in the carpenter's soul. And the form, which is revealed to the carpenter and occurred clearly in his imagination by what the geometer has told him, is as the child to whom his understanding has given birth. Then the carpenter's work is a father with regard to timber, which is the mother of carpentry; and by means

> of the instrument the marriage and ejaculation of sperms occur, which is the effect of every hit by the hammer or cut by the saw, and every cut, separation, and union of the articulated pieces in order to compose forms.[155]

Since knowledge has such a determining role in the making of things, it is important to ask what principles direct knowledge itself. "The nobility of knowledge," says al-Ghazālī, "is in accordance with the nobility of the thing known, and the rank of the knower corresponds to the rank of the knowledge."[156] In a theocentric society like that of premodern Islam, religious principles naturally occupy the highest rank. The Sufis by the very nature of their devotional stance were, of course, at the forefront in calling for the upholding of the religious principles. For them all knowledge stems from God and should therefore be directed toward him. The ultimate aim of knowledge is God, and human sciences were viewed as so many paths leading to that. Al-Ghazālī explains:

> There is no doubt that the most excellent of things known, and the most glorious, and the highest of them, and the most honored, is God the maker (*ṣāniʿ*), the creator, the truth, the one. For knowledge of him, which is knowledge of divine unity, is the most excellent branch of knowledge, the most glorious and most perfect. This knowledge is necessary and must be acquired by all rational beings . . . But this science, though it is excellent in essence and perfect in itself, does not do away with other sciences; indeed, it is not attained except by means of many antecedents, and those antecedents cannot be ordered aright except through various sciences, such as the science of the heavenly bodies and the spheres and the science of all made things. And from the science of divine unity derives other branches of science.[157]

In that manifestation is viewed to have occurred in stages of progressional differentiation, knowledge is also conceived as having degrees that correspond with the universal order of things. The Ikhwān extend the hierarchy of knowledge up to its source, the Universal Soul. They posit that the souls of those who know *(ʿulamāʾ)* know in actuality, while the souls of those who are seeking knowledge *(mutaʿallimūn)* know in potentiality. The act of teaching, then, becomes the bringing out what is in potency into act. This relationship between the souls of the teachers and the pupils reflects that which exists between the Universal Soul and individual souls: the Universal Soul knows in actuality, and individual souls know in potentiality. "Thus, the more knowledge an individual soul has, and the more perfect its products are, the closer it is to the Universal Soul. For thus an individual soul becomes comparable to, and almost identical with, the Universal Soul."[158]

This understanding directs our attention to the criteria of perfection. From the Ikhwān's perspective, perfection can be determined by measuring up the qualities of an artefact produced by an individual soul against the qualities of natural artefacts produced by the Universal Soul. The Ikhwān provide a means to achieve this perfection through the concept of necessary and complementary forms, already discussed. They say that among the innumerable complementary forms, which are responsible for bringing an artefact into its most perfect and noble state, are configuration, movement, light, and purity. Configuration is triangularity, rectangularity, circularity, and similar geometrical forms; movement is any of its six kinds,[159] one of which is translocation *(al-naqla)*, which is of two kinds, circular and rectilinear; and light is of two kinds, essential *(dhātī)* and accidental *(ʿaraḍī)*. Of these complementary forms, the Ikhwān explain, "The noblest of all configurations is the spherical . . . the most perfect of all movements is the circular . . . the most splendid of all lights is the essential, and the purest of all attributes is the transparent."[160] Clearly, the Ikhwān's criteria derive from the world as the noblest and most perfect natural artefact. They say that the body of the world is spherical; the movement of its planets is circular, the light of its stars, except for the moon, is essential; and the quality of its spheres, apart from earth, is transparency.[161] Moving beyond nature up in the universal hierarchy, we find that nature was viewed to operate according to the divine knowledge, and its artefacts reflect the pattern of the higher divine realities. When a human artefact reflects or corresponds to the qualities of natural artefact not only would it resonate with the universal order, but also the maker would be measuring up his work against the work of divinity.[162]

Divine Paradigms

Religious piety advocated by Sufis as well as other religious authorities encourages individuals to become God-like, to acquire the divine character according to human capacity.[163] An often quoted *ḥadīth* says to "acquire in yourselves the character of God."[164] The parameters within which this may be achieved are the divine names, of which the Quran lists the ninety-nine "most beautiful" ones *(al-asmāʾ al-ḥusnā)* and on which there exists a rich medieval genre. The Quran says: "To God belongs the names most beautiful; so call him by them" (7:180). The Prophet is reported to have said: "God has ninety-nine names, whoever recounts them enters paradise."[165] These divine names act as ideals for individuals, for, as Shehadi explains, "If God can be called by such and such names, and man can be called by the same names, and if names are attributive, then one can depict for man the ideal character of God in terms familiar to him. Then man can be enjoined to emulate that character."[166] In this context, human achievements are at their best when they are in the likeness of the divine's.

The divine names were differentiated in premodern Islamic theology into three types: names of essence *(asmā' al-dhāt)*, names of actions *(asmā' al-af'āl)*, and names of attributes *(asmā' al-ṣifāt)*. Names of essence, such as the One *(aḥad)*, the Truth *(ḥaqq)*, concern God alone without regard to the created world. Names of actions, such as Form Giver *(muṣawwir)* and Creator *(khāliq)*, concern God's relations to the created word. Names of attributes, such as the Generous *(karīm)*, the Living *(ḥayy)*, and the Powerful *(qadīr)* concern the qualities of the essence that mediates between the other two. In *al-Maqṣad al-Asnā* al-Ghazālī defines the meaning of each name, explaining to what extent an individual can participate in, or has the capacity to practice, the activities involved. He regards the names as constituting a norm that, on the one hand, defines the nature and the qualitative standard of all human activities and, on the other hand, establishes parameters for human morality and aspirations. The perfection and happiness of humans, he says, lay in their adopting God's characteristics and in embellishing their souls with the meanings of God's names and attributes. A parallel to this view can be found in the Ikhwān's articulation of the notion of art. They say that it is commonly recognized that God loves the skillful, diligent artisan, as described in the *ḥadīth* that says: "God loves the artisan who seeks perfection in his art."[167] Perfection, skill, and diligence in any kind of art, be it intellectual or practical, is to be "in the likeness of the wise artificer, who is God."[168] It is with reference to this, the Ikhwān say, that philosophy is defined as "the similitude of man with God according to human capacity."[169] To emulate God's work in the practice of art is to imitate the patterns and qualities of the divine artefacts, for he is the best artificer, the one who knows best, the wisest, and the noblest. By imitating God's work, the Ikhwān add, one would be attempting to draw nearer to him, referring to the Quranic description of the attitude of the closest angels: "Those unto whom they cry, seeking the way of approach *(al-wasīla)* to their Lord, which of them shall be the nearest" (17:57). To "seek the way of approach" is interpreted as attempting to be as similar to God as human capacity permits through participation in his attributes, according to the Quranic view that one "has only that for which he makes an effort, and that his effort will be seen" (53:39–40).

Although it is not difficult, thanks to al-Ghazālī's examples, to know which of the divine names are directly relevant to the making of architecture, it is not so easy to determine whether these examples relate to actual practices. Yet, even if al-Ghazālī was only using architectural examples in order to illustrate theological concepts, this can still be instructive with regard to the religious framework within which the making of architecture was perceived in premodern Islam. Had the analogy been invalid it would not have been so frequently used.

In the analogy quoted at the opening, al-Ghazālī focuses on the nature of the creative act. The human act of bringing a building into existence, he observes, involves activities similar to those involved in the divine act of creating the world. Both involve the production of an *exemplar* first according to which the object is then made. In drawing the form of a house in whiteness an architect imitates the divine process of manifesting the forms of the world in the *materia prima*. The whiteness of a sheet of paper, in so far as it is capable of receiving an indefinitude of images, is analogous to the *materia prima* wherein, prior to the admission of formal differentiation, all possibilities were fused as a nondifferentiated totality. Ibn ʿArabī also compares this process to "the casting of plaster by a builder in order to manifest in it whatever shapes and forms he wants."[170]

Shifting the focus to the process of production, al-Ghazālī identifies three necessary actions for bringing an object from nonexistence to existence. These are: "designing" *(taqdīr)*,[171] "bringing into existence" *(ījād)*, and "form giving" *(taṣwīr)*. Designing involves both determination and measure; bringing into existence involves production; and form giving involves forming the produced object in accordance with its design. These three activities relate to three consecutive divine names, the Creator *(al-khāliq)*, the Producer *(al-bāriʾ)*,[172] and the Form Giver *(al-muṣawwir)*. Al-Ghazālī writes:

> *Al-khāliq, al-bāriʾ, al-muṣawwir*: it might be thought that these are synonymous names and that they all refer back to creation and innovation, but this is not so. Whatever is brought out from nonexistence to existence requires, first, design *(taqdīr)*, second, bringing into existence *(ījād)* in accordance with the design, and third, form giving *(taṣwīr)* after being brought into existence. God, praised and exalted be he, is a creator *(khāliq)* in that he is a designer *(muqaddir)*; a producer *(bāriʾ)* in that he is an inventor *(mukhtariʿ)*, able to bring things into existence *(mūjid)*; and a form giver *(muṣawwir)* in that he arranges the forms of his inventions in the best *(aḥsan)* order.[173] This is like a building, which requires a designer *(muqaddir)* to determine what is needed in the way of timber, mud-bricks, area of land, the number of stories and their length and breadth. This is normally undertaken by an architect *(muhandis)*, who forms and draws the building. Then a builder *(bannāʾ)* is needed to undertake the works whereby the building fundamentals occur. Then an adorner *(muzayyin)* is needed to chisel the surfaces and adorn its form. This is normally undertaken by a person other than the builder. This is the custom in designing, building and form giving.[174]

Designing, producing, and form giving can thus be seen as forming the core activities of the creative process, be it divine or human. The analogy is not perfect, however. Human limitations introduce two main differences. First, the completion of this creative process requires three different people, whereas God is at once "the designer, the one who brings into existence and the adorner, for he is the creator, the producer and the form giver."[175] Second, unlike the world the finished building can exist independently of its makers. The world, by contrast, is forever dependent on its creator in the manner of speech, which ceases to exist when speaking stops.

Al-Ghazālī's analogy provides insights into the nature of ornamentation in Islamic architecture. From al-Ghazālī's perspective, ornamentation forms an integral part of buildings and not merely superficial decoration. Identifying the name Form Giver with both the order of the universe and the ornamentation of buildings implies that an unornamented building could have been perceived as unordered or unfinished. In this sense, ornamentation is something that is normally thought of in the first stage of the building process, the stage of *taqdīr*, since form giving is only meant to bring out what has already been predetermined in the design.

Al-Ghazālī's analogy also sheds some light on the general absence of human representations in Islamic art and architecture. Viewing the act of form giving *(taṣwīr)*, which includes drawing, painting, and modeling, as a participation in the divine name Form Giver leads naturally to considering other aspects of the creative act, like life giving. To depict the form of a living creature, such as humans and animals, is to make it incumbent upon the depicter to reproduce the living qualities of the creatures they depict. Being incapable of doing so, they would fall short of completing and perfecting their work. This is the line of reasoning presented by many Muslim scholars. Ibn ʿArabī explains why the representation of animated creatures is not desirable while alluding at same time to other preferences.

> The form giver *(al-muṣawwir)* is a man who goes on creating creatures that look like God's creatures, while he is not a creator. Yet, he is a creator, because God said: "create from clay the likeness of a bird" (5:110). He has called him "creator" even though his creature has only the outward figure of a bird. The outward figure of the bird is its form *(ṣūra)*, and every form is capable of manifesting sensible life. God condemned and threatened the form giver because he has not completed the formation of his created form. For it is part of the completeness of its formation that its living nature should be manifest to the senses, and he has no power to do so. By contrast, the depiction of things whose living nature is not manifest to the sense, such as the plant, the mineral, the form of a celestial sphere, and various other configurations, is not like this.[176]

Preoccupation with the precise definition of the human shares in the divine attributes, particularly those concerned with the creative act, must have influenced the ways in which premodern Muslims viewed and rationalized their own actions. As far as the names *khāliq* (Creator) and *bāriʾ* (Animator) are concerned, al-Ghazālī explains, the human share is simply nil. Humans have no access to these names except in a metaphorical sense. But so far as *al-muṣawwir* (Form Giver) is concerned, humans can claim a considerable share: "The servant's share of this name is to acquire in his soul the image of being *(ṣūrat al-wujūd)* in its entirety according to its form and order, until he grasps the form and order of the whole world as though he is gazing at it. From the whole he then descends to the details."[177]

The details include knowing the order of the sensible worlds—man, animal, plant, and mineral—as well as the order of the intelligible worlds—the spiritual and the angelic. Participation in the divine name Form Giver, as al-Ghazālī puts it, points to the ultimate aim of the human creative endeavor:

> This is the servant's share of this name: the acquisition of the intellectual form *(al-ṣūra al-ʿilmiyya)* that conforms with the existential form *(al-ṣūra al-wujūdiyya)*. For knowledge is a form in the soul congruent with the known. God's knowledge of forms is the cause of their existence as external essences, and the forms that manifest these essences are the cause of the existence of the intellectual forms in man's heart. Through this the servant benefits by acquiring knowledge through the meaning of the name Form Giver . . . thereby becoming, by virtue of acquiring this form in his soul, as though he is a form giver.[178]

Chapter 2

Metaphysical Order

Being and Presence

A form giver, al-Ghazālī tells us, is one who has conceived of *ṣūrat al-wujūd* in its entirety. He would have comprehended well the form and order of the universe so that its image becomes vividly present in his soul as though he is constantly observing it. Once this is achieved one descends down to the details.[1] *Ṣūrat al-wujūd*, the "image of being," externalizes al-Ghazālī's *nuskhat al-ʿālam*, the "blueprint of the world" referred to earlier, materializing it in a visible form. Considering the instantaneous nature of the creative act, *nuskhat al-ʿālam* and *ṣūrat al-wujūd* should be understood as separated only by a conceptual distance. Here we will begin to explore *ṣūrat al-wujūd* in both its holistic form and underlying order, focusing on God, Man, and the Word, that is, the *creator*, the *idea*, and the *creative tool.*

Presence and Absence

When dealing with the Sufi conceptions of reality, physical or metaphysical, it is important to recognize the essential difference between their approach and the Cartesian view that conditions our modern understanding.[2] Sufis do not see the world through the Cartesian polarity of subject and object, mind and extension, conscious soul and extended bodies. In fact the subject-object polarity finds neither linguistic nor conceptual support in Arabic. Instead, Sufis present an understanding of the world through the polarity of presence *(ḥuḍūr)* and absence *(ghiyāb)*. Every existent has a presence that matches its mode of being. Even nonexistence has a notional presence. The Quranic polarity of the seen *(shahāda)* and the unseen *(ghayb)* is but an expression of presence and absence. The notion of presence refers to the complex web of physical, mental, and spiritual relationships a being spawns by its very existence and the influences it exerts through this web of connectedness. A thing is perceived to have a presence insofar as it impacts other presences, influences their course of existence, and becomes part of their world. In other words, it is not the mere existence of the thing that matters but rather its level of impact and domain of influence. This is what makes it *effectively* present. Absence is the lack of such efficacy despite physical existence. From the human perspective, what matters is not what exists *out there* but what has a presence in, and an impact on, one's world.[3] In formal correspondences, it is still a common practice in the Arab world today to address the presence of the addressee. In the following, I will be introducing four presences: the primordial, divine, and human presences and the presence of the word. In this context, a "presence" *(ḥaḍra)* refers to a modality of being with all the realities it entails, the relations it involves, and the influences it commands.

The Presence of Geometry

In Sufi metaphysics, numbers and geometry are indispensable tools that aid the reflection on the nature of divinity and illustrate the order of being. Within the bounds of the Euclidean tradition, geometrical principles, such as the point, the line, and the circle, were consistently used to reason about metaphysical realities. As early as al-Ḥallāj (d. 922) we can trace the agency of geometry as a sophisticated hermeneutical tool. Geometrical principles, as will be shown here, are employed to illustrate the initial stages in the creative process, which coincide with the states of universal manifestation. These are the states through which God becomes manifest in various modalities, corresponding to the various levels of differentiation in his creative act. Familiarity with the basic principles of premodern Islamic geometry is therefore necessary, not only to appreciate the agency of geometry in metaphysical sciences but also to gain insights into the spatial reasoning in premodern Islam.

Following the Greek model, premodern Muslim scientists considered geometry to be a part of the mathematical science, which comprises four divisions: the science of number, whose principle is unity or the number one; the science of geometry, whose principle is the point; the science of astronomy, whose principle is the movement of the sun; and the science of music, whose principle is proportion or the equality of two ratios.[4] Along with this structure, geometry contained profound mystical dimensions that survived from the times of its Greek masters, such as Pythagoras, Nicomachus, Euclid, and Plato. The philosophical distinction between the sensible and the intelligible that underpinned the hierarchy of the world was extended to geometry. In their *Rasā'il*, the Ikhwān defined geometry as the science that deals with "measures" *(maqādīr)* and "dimensions" *(ab'ād)*, in their quantitative and qualitative aspects. Distinguishing two kinds of geometry, they wrote: "Know, Oh brother . . . that the study of sensible geometry *(al-handasa al-ḥissiyya)* leads to proficiency in all the practical arts, while the study of intelligible geometry *(al-handasa al-'aqliyya)* leads to proficiency in the intellectual arts, because this science is one of the gates that leads to knowing the substance of the soul, which is the root of all sciences, the element of wisdom, and the origin of all intellectual and practical arts."[5] Sensible geometry was described as the science that deals with sensible measures and configurations, those that can be sensed by sight and touch, whereas intelligible geometry was seen as the science that deals with abstract, immaterial concepts, those that can only be known and understood by the intellect.[6] By virtue of its intellectual nature, intelligible geometry was considered to be the foundation of designing. "When, in his craft, an artisan designs *(qaddara)* before commencing work," the Ikhwān write, "this act involves a kind of intelligible geometry."[7] This shows the utility of geometry as a design tool. Designing, the Ikhwān explain, involves dealing with measures, which are of three kinds: lines, planes, and bodies. These three sensible measures can be conceived of mentally by the qualities of length, breadth, and depth. Length is the intelligible quality of the one-dimensional line; length and breadth are the intelligible qualities of the two-dimensional plane; and length, breadth, and depth are the intelligible qualities of the three-dimensional body. All measures in space can be known by means of these three intelligible qualities. "It is a part of the art of the erudite thinkers *(al-muḥaqqiqīn)*," the Ikhwān say, "to contemplate these dimensions divested of bodies."[8] Geometry was also conceived as being based on the imaginary movement of the point, its generative principle. I will return to this later on. The point was seen as a geometrical reflection of the number 1, with both sharing the same ontological condition. The point was viewed as the principle of dimension, while itself having no dimension, just as the number 1 was seen as the principle of numbers, while itself not being a number. As generative principles both were seen to transcend the domain they manifest, disclosing a mode of reasoning that plays a central role in metaphysical reflections.

Unity of Being

In religious thought, the relationship between the creator and the creatures, God and the world, has always been a central theme. The perplexing questions of why and how God created the world, and what was he doing before creation, have engaged the religious imagination throughout history. The debate on these issues often leads to a sharp distinction between two modalities of being; one belongs to God, the other to the world. In Islam, this debate has continued to unfold a diversity of positions, ranging between the most hermeneutical and the most literal. In this debate the Sufis advocate the doctrine of the Unity of Being *(waḥdat al-wujūd)*. Often misunderstood as blurring the boundary between God and the world, this doctrine emphasizes that there is only one modality of Being *(wujūd)* and that Being proper is none other than God in his most transcendental state. Everything else depends in their existence on this Being who is externalized in many colorful manifestations. As al-Nābulusī reflects: "Being is me, while a being is other than me, because beings *are* by me and I *am* by my Self."[9]

The philosophical reasoning behind this complex concept is rather simple. If God in his primordial presence, before the creation of the world, absurd as this premise may be, is necessarily conceived of as *mawjūd* (is), then *al-wujūd* (being) must either be identical with or other than himself. Otherness implies duality that contradicts the Islamic doctrine of unity, therefore, *al-wujūd* and God must be one.[10] This is only the starting point though, and more complex reasoning is involved when the creation is taken into account. Some important linguistic issues must be considered when dealing with *waḥadt al-wujūd* in English. To begin with, the polarity of "being" and "existence" or "Being" and "being" has no linguistic support in Arabic.[11] There is only one term *wujūd*, with no lower and upper case, that accounts for all shades of meanings and conceptual nuances. The etymological root of this term is *wajada*, "to find," whose passive form, "to be found," means "to exist," and the derivatives *awjada* and *ījād* mean "to bring into existence" and "bringing into existence" respectively.[12] In the Latin-based languages, where the linguistic and conceptual distinction between Being and being, on the one hand, and Being and existence, on the other, has both historical and philosophical depths, the concept of *waḥdat al-wujūd* appears confusing and loses much of its immediacy and transparency. This is reflected in the range of translations available: "unity/unicity/oneness of being/Being" and "unity/oneness of existence/Existence," with all being unsatisfactory equivalents for tending to emphasize either the divine or the worldly side of things, whereas the Sufi concept is clearly about the oneness of both. For simplicity and want of a better expression, I have used the term *Unity of Being*.

In *al-Tuhfa al-Mursala,* "a brief tract on illustrating the science of truths" that entertained wide popularity, Faḍl Allāh al-Barhanbūrī al-Hindī (d. 1620) explains *waḥdat al-wujūd* succinctly:

> Know, O brothers, may God bring you and us happiness, that God *(al-ḥaqq)*, praised and exalted, is Being *(al-wujūd)*; and that this Being has neither shape nor

> limit nor confinement. Yet, in spite of that it has manifested and appeared in shape and limit but without changing from what it was, having no shape and limit: it is [now] even as it was. And [know] that Being is one, while the "clothings" *(albās)* are many and different; that this Being is the reality and inner essence of all beings *(mawjūdāt)*; that all beings *(kā'ināt)*, even the atom, are not devoid of it; and that this Being is not understood in the infinitive sense of realization or happening (suggesting a duality of a subject and a state), for there are no two kinds of Being in the external world. Yet Being cannot be applied to external beings in the same sense as it applies to God, transcendent be he above that. Rather, by Being we mean the Real who is qualified by these qualities, I mean, its self-existence, the existence of everything else by it, and the absence of any other externally. And know that in respect of its inner reality *(kunh)* this Being cannot be revealed to anyone, nor can the intellect, the imagination or the senses conceive of it, nor can it be grasped by analogy. For all these are novelties, transcendent be his essence and attributes above that. And whoever wants to know God in this respect and goes after it is wasting his time.[13]

Thus understood, *waḥdat al-wujūd* sees God as the inner reality of all beings. The oneness of the inner reality in relation to the many and different manifestations is often explained analogically by reference to natural phenomena, such as the invisible, colorless light and its visible, colorful refractions. The concept is usually traced in numerous Quranic verses, such as, "wheresoever you turn, there is the Face of God" (2:115); "We are nearer to him than his jugular vein" (50:16); "And in yourselves. Can you not see?" (51:21); "We are nearer unto him than you are, but you cannot see" (56:85); and "He is with you wheresoever you may be" (57:4).

The Unity of Being becomes a contentious concept once the created world is entered into the equation. Conditions of worldly existence—space, time, and change—cannot be said to apply to the divine, hence the need to distinguish two modes of being, one of Being, the other of becoming, as in the Platonic-Aristotelian metaphysics. According to *waḥdat al-wujūd*, however, there is no need for such a distinction, for the differentiation occurs in the states or modalities that Being takes on at different levels of manifestation. Such differentiation, however, remains external, as the one and same Reality always resides at the very core of things. Viewing *waḥdat al-wujūd* from the point of view of the distinction between Being and becoming, an unnecessary discontinuity between the divine and human modes of existence confuses the concept. While Being may still be seen as the inner reality of all existents, God remains distinct from everything else in the realm of existence. This is often the cause of misunderstanding.[14]

Waḥdat al-wujūd is commonly attributed to Ibn ʿArabī, who articulated it as a central theme in his writings and teachings. Its historical origin, however, cannot be so clearly defined, for there are many Sufi texts from the earliest period of al-Baṣrī, al-Kharrāz, and al-Ḥallāj that clearly express similar conception, though not at the same level of sophistication. ʿAfīfī, in his introduction to Ibn ʿArabī's most famous text *Fuṣūṣ al-Ḥikam*, stresses that the radical doctrine of *waḥdat al-wujūd* is entirely the fabrication of Ibn ʿArabī's imagination and that earlier Sufis had nothing to do with it. In the best light, ʿAfīfī's remark reveals the stance of contemporary Islamic orthodoxy that is eager to show Ibn ʿArabī as an anomaly in an otherwise well-respected, orthodox Sufi tradition. Such a view fails to see the historical depth of the concept, to understand the true nature of the doctrine, and to discern the intrinsic link between the concept itself and the nature of the mystical experience.

Sufi literature presents ample references to *waḥdat al-wujūd*. The ninth-century Sufi al-Kharrāz is often reported as quoting the Prophet's companion ʿUbayda (d. 639), saying: "I have never looked at a single thing without God being nearer to me than it."[15] In *Mishkāt al-Anwār* al-Ghazālī wrote: "The Truth of all truths: from here the Gnostics rise, from the lowlands of metaphor to the peak of the truth; and at the fulfillment of their ascent they see directly face to face that there is naught in existence save only God." He adds, "Each thing has two faces, a face of its own, and a face of its Lord; in respect of its own face it is nothingness, and in respect of the Face of God it is Being. Thus there is nothing in existence save only God and his Face."[16] This is similar to what ʿAbd al-Karīm Al-Jīlī wrote three centuries later: "Being has two aspects: one is pure Being, which is the Essence of the creator, the other is [relative] being associated with nothingness, which is the essence of the creatures."[17] This conception resonates with Ibn ʿArabī's teachings: "In relation to the forms of the world, 'everything will perish,' but in relation to its realities, the world will not perish, nor is it possible to perish."[18] There is no doubt that since Ibn ʿArabī *waḥdat al-wujūd* has become a foundational concept in Sufi metaphysics. As a reflection of an intense spiritual love of God and a burning desire to know him, the doctrine of the Unity of Being can also be seen as an outward expression of the mystical experience. If the affirmation of the absolute unity of God is the cornerstone of the Islamic religion, then the concept of *waḥdat al-wujūd* can be said to be its mystical expression.

The States of Being

While Being in its highest state is identical with the unknowable Essence, this does not mean that it is totally inaccessible by the human mind. Al-Hindī explains that Being has other states whereby it can be known. These states represent degrees of qualification or definition, referred to as "essential determination" *(al-taʿayyun al-dhātī)*, meaning the knowable conditions whereby the divine Essence becomes determined. These states also correspond to the modalities of "universal manifestation" *(al-ẓuhūr al-kullī)*, meaning the knowable conditions whereby the world becomes determined. The hierarchy of these states can be considered from a number of standpoints; however, they are often categorized into seven levels. In *al-Tuḥfa al-Mursala* al-Hindī defines the seven states of Being as follows:[19]

1. Transcendent Unity *(al-aḥadiyya)*: the state of "nondetermination" *(al-lātaʿayyun)*, "absoluteness" *(al-iṭlāq)* and "pure Essence" *(al-dhāt al-baḥt)*, not meaning "that the limits of absoluteness and negation are affirmative in this state, but that *al-wujūd* in this state transcends the addition of qualities and attributes, and is too sacred to be defined by any limit, even the limit of absoluteness."[20] It is the state of the ineffable Essence that "refuses human understanding"; the state of "nonqualified" and "nondetermined" existence that lies beyond human conception. As such, it cannot be the object of any distinctive knowledge, and is, therefore, inaccessible to the human mind.[21]
2. Divine Solitude *(al-waḥda)*: the state of "first determination" *(al-taʿayyun al-awwal)* that represents God's knowledge of his Essence, attributes, and all existents *(mawjūdāt)* in their nondifferentiated, indistinctive mode of being. Ontologically, it mediates between *al-aḥadiyya* and *al-waḥdāniyya* and is also referred to as the state of "Muḥammadan Reality" *(al-ḥaqīqa al-muḥammadiyya)*.[22]
3. Divine Uniqueness *(al-waḥdāniyya)*: the state of "second determination" *(al-taʿayyun al-thānī)* that represents God's knowledge of his Essence, attributes, and all existents in their differentiated, distinctive mode of being. It is also referred to as the state of "Human Reality" *(al-ḥaqīqa al-insāniyya)*. The first three states of Being concern Being in the primordial stage, yet, primordiality as well as precedency and succession in the above states must be understood as intellectual and not temporal qualifications.[23]
4. The World of Spirits *(ʿālam al-arwāḥ)*: the state of simple, abstract cosmic entities, those "in the likeness of which, and in accordance with whose essences, manifestation is fashioned."[24]
5. The World of Similitude *(ʿālam al-mithāl)*: the state of subtle, composed cosmic entities, those which are not susceptible to division, portioning, separation, or conjunction.[25]

6. The World of Bodies *(ʿālam al-ajsām)*: the state of dense, composed cosmic entities, those which are susceptible to division, portioning, separation, and conjunction.[26]
7. Man *(al-insān)*: the last and the sum total of all manifest states, the bodily and the spiritual as well as the states of divine uniqueness and divine solitude.[27]

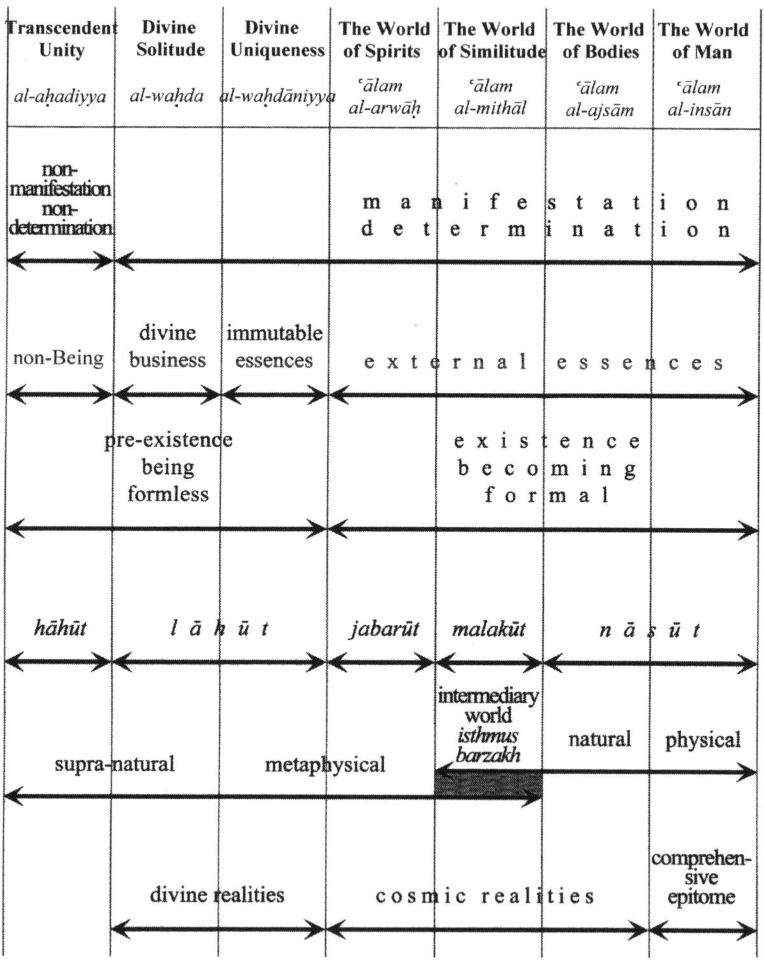

Fig. 2.1 The seven states of Being viewed from various perspectives.

Viewing the states of Being from different standpoints, there is only one state of nonmanifestation *(al-lāẓuhūr)*, that of Transcendent Unity, and six states of universal manifestation.[28] The first three represent the states of Being in the Platonic sense, while the other four represent the states of becoming. In terms of the creative process, there are internal and external modes of differentiation. Internally, the second state is the state of first determination, while the third is the state of second determination. The other four states are concerned with external existence.

These states also identify, as al-Hindī puts it, three distinct modalities or "homes" (*mawāṭin*) of the world. The first home is in the first determination wherein the world is designated as divine "business" *(shuʾūn),* in the sense that the world is God's own concern: "He creates what he will" (30:54).[29] The expression comes from the Quranic verse "every day he is engaged in a divine business *(shaʾn)*" (55:29). The second home is in the second determination, wherein the world is designated as "immutable essences" *(aʿyān thābita)*. The third home is in the "exterior" *(al-khārij)*, in the realm of existence, wherein the world is designated as "external essences" *(aʿyān khārijiyya)*.[30]

In terms of al-Ghazālī's analogies, the three *mawāṭin* of the world can be compared to the stages of architectural production. The first two modalities, the divine business and immutable essence, correspond to the stages of design *(taqdīr)*, wherein the designed object is still contained in an unmanifest mode within the designer's mind. The architectural drawing, which equates al-Ghazālī's *nuskhat al-ʿālam*, the blueprint of the world, corresponds to the world of spirits, the first state of the external essences *(aʿyān khārijiyya)*, which includes both the blueprint and its embodiment. Al-Hindī's hierarchy of the three homes forms the basis of the following analyses of the process of universal manifestation.

The Primordial Presence

The primordial presence *(al-ḥaḍra al-qadīma)*, as distinct from the divine presence *(al-ḥaḍra al-ilāhiyya)*, is the presence of divinity that precedes conceptually all qualifications and determinations, including those of firstness, absoluteness, and unity. It is the presence associated with the first state of Being *(al-aḥadiyya)* from the verse: "Say: He is God, the One *(aḥad)*!" (112:1). Although it has a conceptual presence, this state is characterized by existential absence, as our comprehension of it is based on the denial of all comprehensible definitions and conditions. It is the presence of the Essence that the Sufis compare to the geometrical point.

Nondetermination

In the *Ṭawāsīn*, al-Ḥallāj refers to the point as "the meaning of unity, but not Unity."[31] He presents several mystical and graphical references to the symbolism of the point, the line, the circle, and the alphabet. "The circle has no door," he writes, "and the point in the middle of the circle is the meaning of truth."[32] In the Sufi teachings, the symbolism of the point *(nuqṭa)*, literally "drop," "dot," is a consistent and recurrent theme. In *Kitāb al-Nuqṭa* (The Book of the Point), al-Jīlī presents a sophisticated exposition of the meaning of the point, seen as a potent symbol of the ultimate Reality *(ḥaqīqat al-ḥaqā'iq)*, a graspable geometrical principle capable of revealing the relationship the divine Essence bears to the world.[33] In their fascinating discourse on the symbolism of the point, Sufis often quote a prophetic tradition that reduces all human knowledge to the dot of the Arabic letter *bā'* (B). I will discuss this later.

To understand the ontological relation Sufis draw between the point and the first state of Being, we need to reinvoke the Ikhwān's distinction between the sensible and intelligible geometry, which they extend to the point. The sensible point *(al-nuqṭa al-ḥissiyya)*, they explain, is a physical entity that has parts, whereas the intelligible point *(al-nuqṭa al-ʿaqliyya)* is a nonspatial principle that has no parts.[34] In a treatise on the alphabetical symbolism, al-Jīlī reiterates this differentiation: "Know that, in reality, the point cannot be determined by sight, because [it is indivisible while] all that is manifested by it in the bodily world is divisible. So the perceived point is an expression *(ʿibāra)* of its reality, the definition of which is a "single, indivisible substance" *(jawhar fard lā-yatajazza')*."[35] Thus understood, the sensible point is the smallest spatial entity in the Euclidian geometry whose repetition produces a line, the repetition of which produces a plane, the repetition of which produces a volume. This spatial entity has a dimension, however indefinitely small. For the repetition of a dimensionless point cannot produce a dimension any more than the addition of zeros can produce a number. By contrast, the intelligible point is a mental concept, denoting the dimensionless, indivisible principle that lies beyond the confines of spatial conditions. The sensible point is the physical embodiment of the intelligible point, which not only escapes space and spatial conditions but also defies our affirmative comprehension. We are unable to attribute to the point any essential qualities whereby it may become affirmatively graspable. When we describe the point as indivisible, formless, dimensionless, without extension or duration, and so forth, we are in fact negating its spatio-temporal qualities. Negative attributes can only tell us what a thing is *not*. Although negating is a form of knowing, an entity can only be grasped by means of its affirmative attributes. As Ibn ʿArabī explains that "negation is not an essential attribute *(ṣifa dhātiyya)*, for all essential attributes of beings are affirmative *(thubūtiyya)*."[36] Furthermore, when we try to define the intelligible point by negative attributes, we are forced to employ spatio-temporal qualities, which the intelligible point transcends. By such definitions we reveal our tendency to presume that such sensible qualities are potentially latent in it. However, neither can these negative attributes determine, limit, or define the quality of the intelligible point nor can they render it graspable.

The ungraspability and incomprehensibility of the point renders it a potent symbol of the ineffable divine Essence *(al-dhāt)* or God in the state of nondetermination. "Whenever I speak of the Point I mean the Secret of the Essence," says the twentieth-century Sufi Aḥmad al-ʿAlawī in his treatise on the symbolism of the Arabic letters.[37] And al-Jīlī says that "the point is a symbol *(ishāra)* of God's essence that is hidden behind the veils of his multiplicity."[38] The point stands for the Essence because it is just as ungraspable and incomprehensible to say that the point is formless, dimensionless, indivisible, and so forth, as the Quranic description of God: "Naught is as his likeness" (42:11). Negation is the only way to know of the divine Essence and, by extension, of its symbol, the point.[39] Ibn ʿArabī explains the state of divinity to which the point is ontologically tied: "Praise be to God before whose oneness there was not a before, unless the Before was He, and after whose singleness there is not an after, except the After be He. He is, and there is with Him no after nor before, nor above nor below, nor far nor near, nor union nor division, nor how nor where nor when, nor time nor moment nor age, nor being nor place. And He is now as He was. He is the One without oneness, and the Single without singleness . . . He is the First without firstness, and the Last without lastness. He is the Outward without outwardness, and the Inward without inwardness."[40]

Sufis teach that the point is the principle of geometry just as 1 is the principle of number. Here the correspondence can be observed on two levels. In Arabic "unity" is denoted by two terms: *aḥad,* as in "Say: he is God, the One *(aḥad)*" (112:1); and *wāḥid*, as in "Your God is One *(wāḥid)* God" (2:163). The conceptual difference between these two terms can be traced in the Sufi teachings. The unmanifest state of Transcendent Unity is designated by the term *al-aḥadiyya*, from *aḥad*, while the two manifested states of the divine Solitude and Uniqueness are designated by the terms *al-waḥda* and *al-waḥdāniyya*, respectively, from *wāḥid*. The two manifest states of Being share the same etymological root with the number 1, *wāḥid*, whereby they are deliberately distinguished from *al-aḥadiyya*, whose root, *aḥad*, though it connotes the idea of unity, signifies unity without likeness, not even in numbers. In geometrical terms, the sensible point, as the principle of sensible geometry, corresponds to the 1, as the principle of numbers. For just as the number 1 is a reflection of the unmanifest Unity *(aḥad)* in the realm of numbers, so the sensible point is a reflection of the intelligible point in the realm of geometry. The unmanifest Unity, however, is that which lies beyond both realms of number and geometry altogether. It can be analogically equated with the 0 or the pure "whiteness" *(bayāḍ)* in Ghazālī's example. Whiteness and 0, as Unity unaffirmed, are symbolic expressions of that which lies beyond the first comprehensible affirmation of Being, expressed by the sensible point and the number 1. Zero is to 1 what the whiteness is to the point and what possibility is to actuality: the state of inconnumerable unity and infinite multiplicity.[41]

In *al-Tuḥfa al-Mursala,* al-Hindī takes this analogy a step further. He says that Being *(wujūd)* is to beings *(mawjūdāt)* as light is to colors and figures. Being is the reality whereby things become conceivable just as light is the condition whereby colors and figures become perceivable. But unlike light, he says, Being in its manifestation is ceaseless and more intense, and, therefore, only the elites are aware of its presence.[42] He further explains that all beings *(mawjūdāt)*, in respect of Being *(wujūd)*, are none other than the Real *(al-ḥaqq)*, but in respect of determination are other than the Real. Otherness is a relative matter. In reality, he adds, even the shape is none other than Real. To illustrate this view, al-Hindī uses the example of the various objects made out of water, such as hail, waves, and ice in the form of a cup. In reality, he says, all of these are none other than water, but in terms of their specific forms they are obviously other than water. Likewise is the mirage, which is air appearing in the form of water; in reality, it is none other than air, but in appearance it is other than air.[43]

Relating this to space, the geometrical point in its two modalities, the sensible and the intelligible, may be taken to represent the ubiquitous presence of Being in both its determined and undetermined states. Viewed from al-Hindī's standpoint, the point can be seen to be the basis of spatial compositions in the same way that Being is considered to be the inner reality of all beings. Ibn ʿArabī develops this argument considering the point in its own right and in what it causes to appear in the form of spatial composition. In reality, Ibn ʿArabī says, a spatial object is none other than the point, but in determination it is other than the point. Explaining the nature of the radius, he writes: "A line terminates at a point. Its beginning and its end are and are not parts of it, as you may wish to say. What should be said of the line is this: neither are the points the line itself nor other than itself . . . The line is made up of points, it cannot be conceived in any other way. The plane is made up of lines, so it is made up of points, and the body is made up of planes, so it is made up of lines, which are made up of points."[44]

In summary, the point, itself undetermined and unmanifest, is the principle of determinate manifestation. It is to space what the divine Essence is to the world: the unmanifest principle of manifestation. In its intelligible mode, it encompasses the entirety of space, for potentially all is conceived within it. In its sensible mode, it is the generative principle of space, for all bodies in space can be geometrically reduced to a point: it is both the whole and the part.

The Divine Presence

The divine presence *(al-ḥaḍra al-ilāhiyya)*, as distinct from the primordial presence *(al-ḥaḍra al-qadīma)*, is the presence wherein God is known through his names and attributes. It is the state in which the unity of the Essence becomes associated with the multiplicity of the names and attributes. The Sufis teach that in his primordial presence God desired to be known, to reveal the mysteries of his inner treasure, so he descended from his incomprehensible supremacy, the state of Transcendent Unity, through the state of Solitude, to the state of Uniqueness. Therein he revealed his names and attributes as means whereby he may become knowable. By this descent the nonqualified and undetermined Being becomes manifest in two qualified and determined states—first and second determination—revealing the divine presence.

Unity and Multiplicity

The act of manifestation is associated with a perplexing philosophical question: how could the simple unity produce the rich and complex multiplicity? The tradition that says, "God was and nothing with him; and he is now even as he was," raises other related questions.[45] How could God remain as he was after creating the world? How could God, the one, when there was nothing with him remain the same one when the multitude of existents is associated with him? This is the paradox of unity and multiplicity implicit in the act of manifestation, the paradox of the one becoming many and at the same time remaining one, of God being at once the name and the named.[46] The key to understanding this paradox, the Sufis teach, is the double negation: to think of external beings as neither God himself nor other than himself. It is like looking in a mirror and seeing your image: the reflected image is neither yourself, since you are standing apart from the mirror, nor other than yourself, since it is your own and not anyone else's. If you imagine that you are able to look in a number of mirrors simultaneously and see your reflected images in all of them at once, then the paradox of unity and multiplicity is partially resolved. And if you imagine these mirrors to be infinite in number, reflecting infinite aspects of your personality, then the paradox is almost resolved. What remains to be explained is the "mirrors": what are they and where do they come from? In this analogy, the mirrors are none other than the created beings, the things of the world, whose appearance coincides which the manifestation of the divine reflections or realities.[47] At the first level of determination, these realities are described as divine business *(shu'ūn ilāhiyya)*, whereas at the second level of determination they are designated as immutable essences *(aʿyān thābita)*.

First Determination

Al-taʿayyun al-awwal, the state of the first determination, al-Hindī explains, designates God's *knowledge* of his Essence, attributes, and all beings *(mawjūdāt)* in an undifferentiated, indistinctive mode. This state is often likened to the existence of a tree in the seed prior to its physical materialization. The act of *knowing* is the first affirmative attribute that determines the Essence. Knowing implies a triplicity: knower, known, and knowledge, affirmatively differentiating among the Essence (knower), the names (known), and the con-

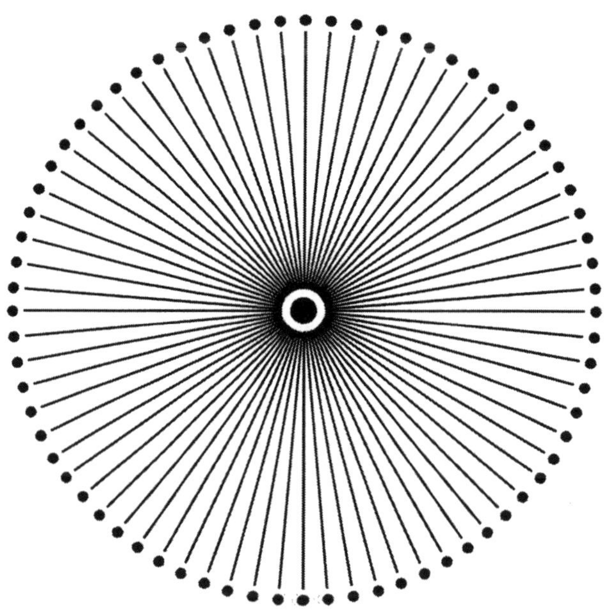

Fig. 2.2 The circle as a symbol of divinity in the state of first determination.

nection (knowledge); or among unity, multiplicity, and affinity. In geometrical terms, this initial order can be traced in the ternary structure of the circle seen as the first qualification of the point: the unity of the center (knower), the multiplicity of the points of the circumference (known), and the connecting radii (knowledge).

Sufis see the revealment of the divine's infinite names from the incomprehensible Essence as analogous to the projection of the circumference's indefinite number of points from the indivisible center and to the reflection of God's "forms" in the mirrors of beings. Through this ontological relationship, the circle becomes the symbol of the first comprehensible form of unity the Essence takes on. The circle's inherent geometrical qualities are thus conditioned by the metaphysical reality it embodies. The circle, therefore, offers effective cues that help us understand the paradox of unity and multiplicity. Ibn ʿArabī explains:

> Every line projecting from the center to the circumference is equal to its companion and terminates at a point on the circumference. In itself the center neither multiplies nor increases despite the multitude of lines that project from it to the circumference. The point of the center relates to every point on the circumference by its same essence. For if it were to relate to one point on the circumference by other than that by which it relates to another, it would be divisible, and it would not be true that it is one, yet it is. So it relates to all the points, in spite of their multitude, by none other than its essence. It is certain then that multiplicity manifests from the one Essence without this Essence being multiplied.[48]

The Quran teaches that the world depends in its existence on God: "O mankind! It is you who are in need of God"; while God is "the self-sufficient, the glorified" (35:15). He is "independent of all creatures" (3:97).[49] Ibn ʿArabī sees in the circle a confirmation of this. Although the center and the circle are mutually dependent on each other's presence, in the sense that circularity demands a center just as centrality demands a domain, the center, as a point, remains autonomous and self-sufficient on its own.[50] The circle, by contrast, has no state wherein it can dispense with its dependency on the center. Just like an image in a mirror: its existence depends upon the presence of the one whose image is being reflected while the one remains independent on its own.

In the state of first determination, al-Hindī explains, the world is designated as divine business *(shuʾūn ilāhiyya)*, a modality that, according to Ibn ʿArabī,

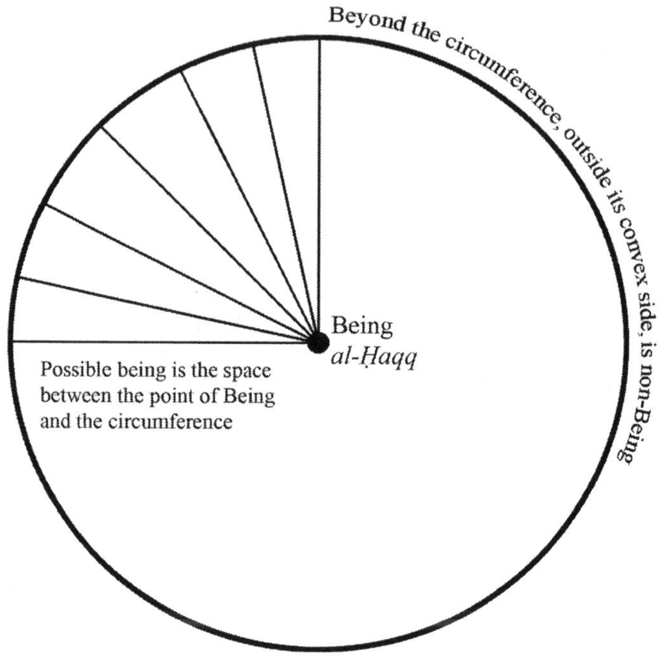

Fig. 2.3 The world as divine business *(sha'n)* according to Ibn ʿArabī *(Futūḥāt)*.

differentiates three things: Being *(al-wujūd)*, non-Being *(al-ʿadam)*, and the possible *(al-mumkin)*. This corresponds to the ternary structure of the knower, known, and knowledge and its geometrization in the circle. Ibn ʿArabī explains: "The divine business *(al-sha'n)* in itself is as the point in relation to the circumference and that which is in between. The point is Being *(al-ḥaqq)*, the space outside the circumference is non-Being *(al-ʿadam)*, or, say, darkness, and that which is in between the point and the space outside the circumference is the possible *(al-mumkin)* . . . We have been given the point because it is the origin of the existence of the circle's circumference that was manifested by the point. Likewise, the possible does not manifest except by Being and the circumference of the circle."[51]

The differentiation of Being, non-Being, and the possible at the first level of determination is also described as God descending from the level of absolute oneness to the level of singleness (*fardiyya,* from *fard,* "odd").[52] Singleness is the level of the affirmative differentiation of the act of knowing, which is coincidental with the creative act. Knowing presupposes the existence of the *known*, the possible world. The nature of the divine creative act is in harmony with the initial structure of determination. Ibn ʿArabī explains:

> Know, may God guide you, that the creative order is in itself based on singleness *(fardiyya),* wherein triplicity *(tathlīth)* is implicit, since singleness begins from the number 3 upward. Three is the first single (*fard*, "odd") number. It is from the presence of singleness that the world has come into existence. God says: "and our word unto a thing, when we intend it, is only that we say unto it: Be! and it is" (16:40). Here is an Essence, one with a Will and a creative Word. Without this Essence, its Will, which is its turning towards bringing something in particular into existence, and its uttering of the word "Be!" to that particular thing at the moment of turning, that thing would not have come into existence.[53]

In the *Futūḥāt* Ibn ʿArabī translates this creative triplicity into a geometrical form. He illustrates the triplicity of the divine creative act with reference to the circle: the center stands for the Essence, the radius for the Will, and the circumference for the coming into being by the word Be!

> The line projecting from the central point to a single point on the circumference represents the predestination each creature has from its creator-most transcendent. It is his saying: "and our word unto a thing, when we intend it, is only that we say unto it: Be! and it is." Will here is that line we assumed as projecting from the point of the circle to the circumference. It is the divine orientation *(al-tawajjuh al-ilāhī)* that determines the existence of that point of the circumference. The circumference is the same "circle of potential beings" *(dāʾirat al-mumkināt),* and the point in the center, which determines the points of the encompassing circle, is the necessary, self-sufficient Being *(wājib al-wujūd li-nafsihi).*[54]

Thus viewed, the circle's inherent order provides an immediate expression or materialization of many Quranic concepts. Referring to the verse, "he is the first and the last, and the outward and the inward" (57:3), Ibn ʿArabī says: "The world is between the center and the circumference: the center is the first, and the circumference is the last."[55] He adds: "every point of the circumference is an end to a line, while the point out of which a line projects to the circumference is the beginning of that line, so he is the first and the last. He is the first of every possible being just as the point is the beginning of every line."[56] And with reference

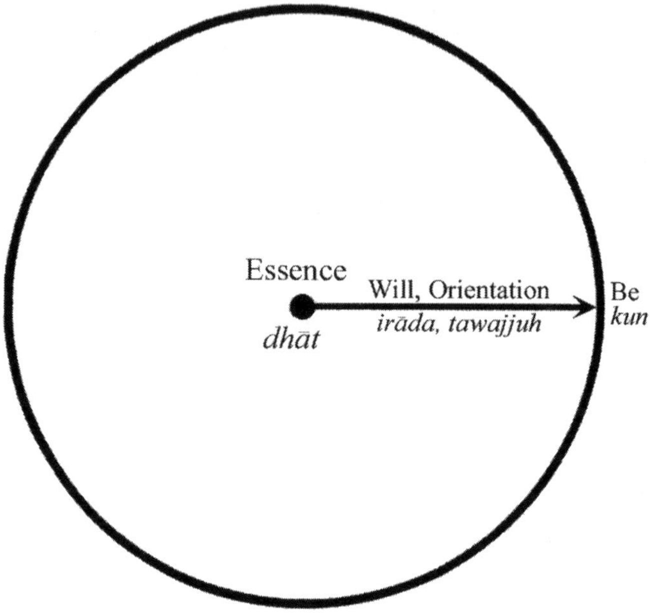

Fig. 2.4 The geometric representation of the divine creative command according to Ibn ʿArabī.

to the verses, "And God, all unseen, surrounds them" (85:20), and "Verily, he is surrounding all things" (41:54), Ibn ʿArabī writes: "If you assume lines projecting from the point to the circumference *(muḥīṭ),* these will terminate but unto a point. The whole circumference bares the same relationship to the point, which is his saying: 'And God, all unseen, surrounds *(muḥīṭ)* them,' and his saying: 'Is not he surrounding *(muḥīṭ)* all things?'"[57]

The Arabic word for "circumference," *muḥīṭ*, which also means "surrounding" and "encompassing," provides a linguistic support for the mystical interpretations.[58] The symbolism of the circle is also used to illustrate and confirm some theological dogmas, such as the ultimate return to God at the end of the world. The lines projecting from the point of the center to the points of the circumference that stand for all possible beings remind us, according to Ibn ʿArabī, of our ultimate destination, as stated in the Quran: "and unto him the whole matter will be returned" (11:123); "As he brought you forth, so you shall return" (7:29); "God initiates the creation, then the recreation, then unto him you will be returned" (30:11).[59] Ibn ʿArabī sees in the circle and the spherical form of the cosmos a constant reminder of this ultimate return.

> Know that since the world is spherical in shape, man longs at the end of his life to his beginning. Our springing forth from nonexistence to existence was by him, and to him we shall return, as he says: "and unto him the whole matter will be returned" . . . Do you not see how when you start drawing a circle you keep encircling the line until it terminates unto its beginning: then it is a circle. Had the matter been otherwise, had we sprung forth from him in a straight line, we would not have returned unto him, and his saying: "then unto him you will be returned" would not have been true, but he is the truthful.[60]

Second Determination

The circle, the symbol of the state of first determination, is the first comprehensible form unity takes on in the process of manifestation. It represents the first polarization that differentiates the unity of the Essence from the multiplicity of the names, but without distinguishing the names from one another. Just like the points of the circumference: they are all alike and equally related to the original point-center. At the state of second determination *(al-taʿayyun al-thānī)*, the divine names become differentiated and distinguished from one another, manifesting an infinite number of patterns. These patterns crystallize the infinite sets of relationships and combinations that bind various divine names together. Geometrically, these patterns can be seen as an indefinite number of geometrical configurations that can be inscribed within a circle, each as it were crystallizing one of the divine patterns. A square, for example, can be seen to geometrically crystallize the pattern of quadrature, which in turn crystallizes the relationship that binds any four distinct yet related divine names, such as the *Living*, the *Knowing*, the *Willing*, and the *Powerful*. Ibn ʿArabī assimilates this process to the differentiation of geometrical forms within a circle: "The world in its entirety is circular in form, within which are then differentiated the forms of all figures, such as quadrature, triplicity, hexad, and so on indefinitely."[61]

For Sufis, combinations of the divine names constitute the regulating patterns of existence, varying according to the subjects they designate. The creation of the world, for example, requires a pattern different to that required for the subsistence of the world after it has been created. Likewise, different modalities of the divine reveal different patterns. A different combination of attributes is needed to know God as the creator of the world than the ones needed to know him as the self-sufficient. Sufis discern a structure in these infinite varieties of patterns, based on a perceived hierarchy in both single and combined divine names. Quadrature and triplicity occupy, with regard to the creative process, primary positions in this hierarchy.

Quadrature: Pattern of Proliferation

Sufis teach that God's manifestation and becoming knowable coincides with the creation of the world. For this creative emergence to be fulfilled a certain combination of divine names is necessary. According to Ibn ʿArabī, this is achieved through four principal names: the *Living (al-ḥayy)*, the *Knowing (al-ʿālim)*, the *Willing (al-murīd)*, and the *Powerful (al-qadīr)*, which manifest the attributes of Life, Knowledge, Will, and Power. There are several Quranic references to these names: "God! There is no God save him, the Alive, the Eternal" (3:2); "He creates what he will. He is the Knower, the Powerful" (30:54). The logic that brings these four names together is based on human creativity that necessitates these four indispensable attributes. In order to be able to produce anything, one must, first of all, be alive, must know what one is intending to produce, must have the will for production, and must be able to produce. This is the creative quadrature according to human logic. It might derive from the physical order, but its roots lie in the metaphysical world. The physical order of existence, as Ibn ʿArabī affirms, is necessarily rooted in divine realities.[62] If this is the way we naturally conceive of the creative act, it is because divine realities structure our modes of thinking according to the original order of things.[63] This creative quadrature reveals the first trace of order, the *exemplar* of all quadratures in the created world. The nearest cognate pattern to the creative quadrature is that which binds the divine names mentioned in the verse: "He is the First and the Last, and the Outward and the Inward" (57:3).[64] Already linked to the structure of the circle, this combination describes the position of the creator in the created world.

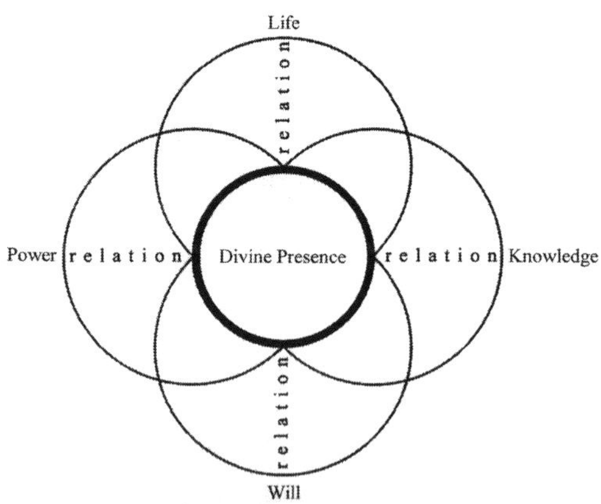

Fig. 2.5 The first stage of manifestation according to Ibn ʿArabī *(Futūḥāt)*.

Within the creative quadrature Ibn ʿArabī traces a hierarchy. Life, he says, is the primary attribute since it is the necessary condition for all other attributes. Knowledge follows Life in the hierarchy. Al-Qāshānī considers Knowledge to be the leader of the attributes, on the basis that although one's creative ability hinges on being alive, being alive does not presuppose the other creative attributes.[65] For Ibn ʿArabī, however, Life is the very condition of existence, including the existence of Knowledge, hence its primacy. Most Sufis, however, concur with Ibn ʿArabī on the primacy of both Life and Knowledge over Will and Power. The logical basis of Ibn ʿArabī's hierarchy concerns the domain of influence and the limits of the realm to which the presence of each of these attributes extends in the creative process. Being the very condition of all attributes Life occupies the highest position in the hierarchy. Knowledge follows, since its domain of influence extends to both the realm of necessary Being and the realm of possibility, which includes what can and cannot exist *(al-mumkin wa al-muḥāl).* Then follows Will, whose domain of influence is restricted to the realm of the possible by exercising a preference as to what may and may not exist. Power is the last in the hierarchy, being the most restricted, since its function is confined to bring into existence the possibility already given a preference for existence.[66]

This hierarchy is conducive to another kind of differentiation, in which each attribute acquires an active or passive generative quality. The primary attributes of Life and Knowledge are considered to be active, acting upon the secondary attributes of Will and Power, which are passive. Ibn ʿArabī considers that just as Life is the condition of Knowledge, Will is the condition of Power. So Will is to Power as Life is to Knowledge. Accordingly, Life draws Will into

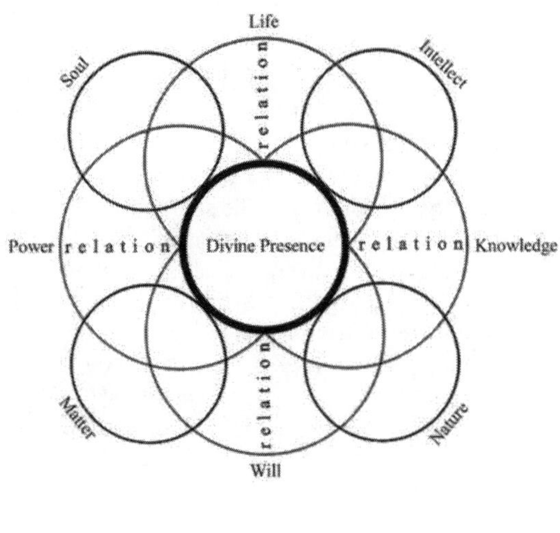

Fig. 2.6 The second stage of manifestation according to Ibn ʿArabī *(Futūḥāt).*

its state, and Knowledge draws Power into its state. The articulation of the creative quadrature's original hierarchy into active and passive, conditioning and conditioned, form the divine model for all other creative orders that rely on the agency of active or passive qualities. "The created world, in its entirety, is passive in relation to God," Ibn ʿArabī says, "while in itself is active in parts and passive in others."[67]

The creative quadrature of divinity is novel since it occurs without a preceding model. Every quadrature that follows, in this as well as in other domains, is but a reproduction of this novel order. With reference to the prophetic tradition, which says that creation started with the Intellect *(al-ʿaql)* followed by the creation of the Soul *(al-nafs)* from the Intellect, Ibn ʿArabī traces the second stage of manifestation: "God brought into existence the first Intellect from the attribute of Life and the Soul from the attribute of Knowledge. So the Intellect was the condition for the existence of the Soul as Life was the condition for the existence of Knowledge. The two passive realities in relation to the Intellect and the Soul were Universal Matter *(al-habāʾ)* and Universal Body. These four were the origin whence all forms in the world were manifested."[68]

At this stage Ibn ʿArabī introduces a new element, "Nature" *(al-ṭabīʿa)*, between the Soul and Universal Matter. Nature, as Ibn ʿArabī describes it, is an intelligible reality, a force, that has no essence *(ʿayn)*. We know it through its effects in the physical world, which manifest through the agency of four realities—heat *(ḥarāra)*, cold *(burūda)*, dryness *(yubūsa)*, and moistness *(ruṭūba)*.[69] In our bodily experiences we sense the effects of these four forces of Nature in the phenomenal world, not Nature per se. The working of Nature, Ibn ʿArabī explains, displays a quaternary structure that resonates with the original creative quadrature:

> Between the Soul and Universal Matter there is the state of Nature. It, too, is based on four realities, two of which are active, and two are passive. Yet all are in the state of passivity with regard to the source whence they proceeded. These are heat, cold, moistness, and dryness. Dryness is passive in relation to heat, and moistness is passive in relation to cold. Heat is from the Intellect, and the Intellect is from Life; hence the nature of life in the sensible bodies is heat. Cold is from the Soul, and the Soul from Knowledge; hence knowledge, when settled, is usually described by the "cold of certainty" and by "snow"[70] . . . As dryness and moistness are passive with regard to heat and cold, Will demands dryness because it belongs to its state, and Power demands moistness because it belongs to its state. And since Power is restricted to bringing-into-existence in particular, it is duly charged with imprinting the nature of life, that is, heat and moistness, in the bodies.[71]

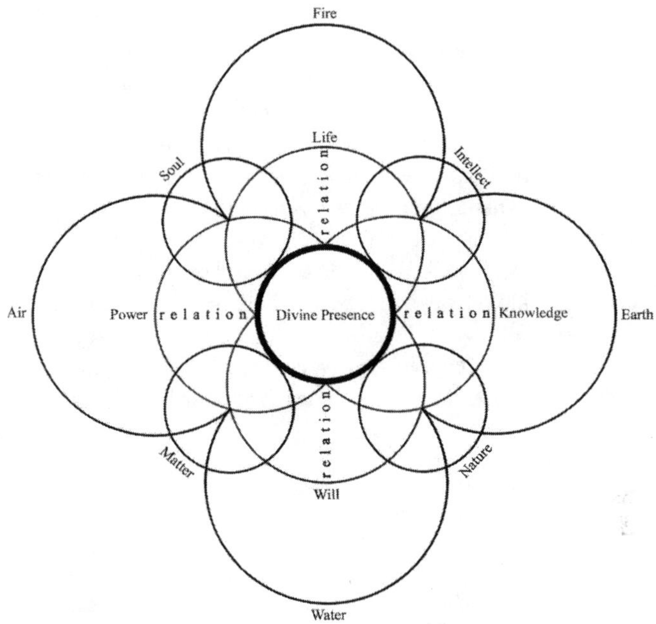

Fig. 2.7a The third stage of manifestation according to Ibn ʿArabī (*Futūḥāt*, a: Dār Ṣādir ed.).

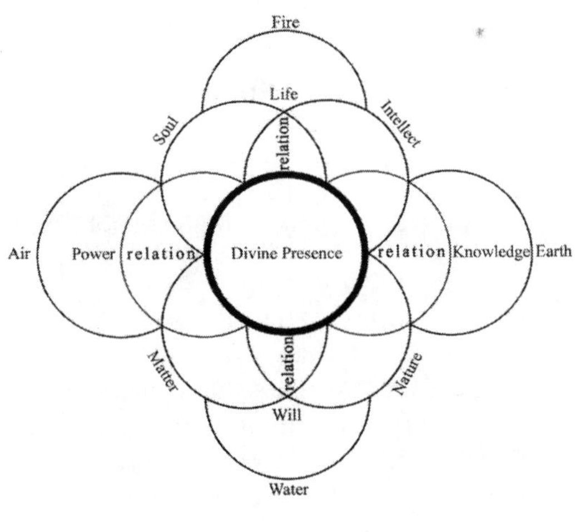

Fig. 2.7b The third stage of manifestation according to to Ibn ʿArabī (MS. 1328).

Ibn ʿArabī says that all forms *(ṣuwar)* and figures *(ashkāl)* are manifested in the Universal Matter and the Universal Body. The form of our world is one of these "forms and figures." Referring to the Quranic verse, "the heavens and the earth were of one piece *(martūqa)*, then we parted them, and we made every living thing of water" (21:30), he says, heaven and earth were first manifested in the form of an undifferentiated, indistinctive mass (*martūqa,* literally "stitched together"). "Then God turned to unstitching the sewn mass in order to distinguish between their (i.e., heaven and earth) essences *(aʿyān)*. Water was the principle of their existence, and that is why he said: We made every living thing of water."[72] This provides the key for the third stage in the creative process, the stage of manifesting the four simple and ideal elements, Fire, Air, Water, and Earth. Water is the principal element, just as Life, to which it relates, is the principal divine attribute. Ibn ʿArabī explains the working of the quaternary order at the third stage of manifestation:

> First of all, God ordered these four natural realities in a particular pattern *(naẓm makhṣūṣ)*. He joined heat and dryness, and there was the simple and ideal Fire. He manifested its ruling *(ḥukm)* in three places of the Throne's body *(al-ʿarsh),* which is the "utmost sphere" *(al-falak al-aqṣā)* and the "Universal Body" *(al-jism al-kull)*. He called the first place "Aries," called the second place, which is the fifth of the designated places, "Leo," and called the third place, which is the ninth of the designated places, "Sagittarius." Then he joined cold and dryness, and there was the simple and ideal Earth. He manifested its ruling in three places of this sphere: He called the first place "Taurus," the second "Virgo," and the third "Capricorn." Then he joined heat and moistness, and there was the simple and ideal Air. He manifested its ruling in three places of this utmost sphere: He called the first place "Gemini," the second "Libra," and the third "Aquarius." Then he joined cold to moistness, and there was the simple and ideal Water. He manifested its ruling in three places of the utmost sphere: He called the first place "Cancer," the second "Scorpio," and the third "Pisces." This is the division of the sphere of the constellations: there are twelve designated divisions determined by the twenty-eight planets. All is set by the design *(taqdīr)* of the Mighty, the Knower.[73]

This pattern of quaternary manifestation also represents the basic pattern of the cosmogonic proliferation of the many from the maternal source, the point. As the unmanifest source that precedes the manifestation of the many, the point signifies the feminine, progenitive origin wherein all things are conceived as an undifferentiated totality. "Whatever may be the term that you do choose for the first entity," Ibn ʿArabī explains, "it will always be feminine."[74] Geometrically, the unity of mother-point proliferates into the multiplicity of the circumference. The points of this circumference represent the genera of

Fig. 2.8 Pattern of proliferation according to Ibn ʿArabī.

existents. In the same manner that the mother-point first burst open to give birth to the first circumference, the multiplicity of genera, so likewise each genus may in turn proliferate into a multiplicity of species, species into kinds, and kinds into individuals.

Ibn ʿArabī says that every part of the world may cause the existence of another smaller world similar to, but in no way more perfect than, it; and so likewise "every point may cause the existence of a circumference whose condition is the same as the first one, and so on ad infinitum."[75] "The principle of all this is the first point," he writes, "for a line extending from the point-center to a determined point on the circumference may also extend from it to the points of the half-circle that lies outside the first circle."[76] The pattern of proliferation at once follows and inscribes the law of unity and multiplicity or the whole and the parts, according to which every part reveals the same order of the whole and as such it forms a whole on its own.

Triplicity: Pattern of Formation

Reflecting on the nature of the creative act, Ibn ʿArabī says that we can only say that God "designs" *(yuqaddir)* things eternally but not "brings into existence" *(yūjid)* eternally, because it is not possible.[77] Accordingly, he distinguishes two modes of creation: creation by "designing" or "determin ation" *(taqdīr)* and creation by "bringing into existence" *(ījād)* or "formation" (*takwīn)*. The former is an eternal, creative act that does not involve physical production, whereas the latter is an act that does. The former coexists with God's knowing of the nonexistent world in its potential state, whereas the latter coexists with the bringing of the world from nonexistence into existence, from potentiality into actuality.[78] The distinction between these two creative modalities is traceable at the linguistic level. In Arabic *ʿālam* and *kawn* are the two distinct terms translated as "world." The former derives from *ʿilm*, "knowledge," with reference to which Ibn ʿArabī says that the moment God knew himself, he knew the world.[79] The latter derives from the trilateral root *k.w.n.,* literally, *ḥadath*, that is, "something new," "a novelty," "an unprecedented thing," "occurrence," from which comes the terms *muḥdath* and *ḥudūth*, "ephemeral existence" and "newness," as opposed to *qadīm* and *qidam,* "primordial" and "eternity."

Although knowing is a creative act, it is not necessarily a physically productive one. Things may be created in the imagination without being brought into existence physically. *Takwīn*, however, is necessarily a physically productive act, causing the designated thing to exist. The verb *kawwana* means "to bring into existence," "to synthesize," the imperative of which *kun* (Be!) is the divine creative word. Thus creation has two complementary modes: *designative* through knowledge and design *(taqdīr)*, and *productive* through bringing into existence and formation *(takwīn)*. This conception further supports al-Ghazālī's analogies already discussed.

If quadrature can be viewed as the primary pattern of creation with regard to *taqdīr*, then triplicity is the primary pattern of creation with regard to *takwīn.* This can be traced in the Sufi conceptions of the creative act and of the divine model for bodily formation. The divine creative command, as we have seen, is based on the triplicity of the Essence, Will, and Word. This is considered to be an active triplicity in response to which a passive triplicity appears in the created thing. It is the union of both that causes this thing to exist. In response to the creative command, Ibn ʿArabī explains,

> there arises in the thing to be created, too, a singleness, a triplicity, by which the thing, on its part, properly partakes in its own formation and its being brought into being. This is its thing-ness *(shayʾiyya)*, its hearing *(samāʿ)*, and its obeying *(imtithāl)* the command of the creator concerning its coming into being. So the thing matches the [creative] triad with a triad: its affirmative, though nonexisting, essence corresponds to the Essence of its creator; its hearing corresponds to the Will of its creator; and its obedient acceptance of what

> has been commanded concerning its formation corresponds to the creator's utterance of Be![80]

Formation *(takwīn)* presupposes embodiment that demands a pattern of divine names to be realized. This pattern determines the spatiality of embodiment through the three dimensions and six directions. Sufis distinguish two complementary modalities whereby God may be known: the creative modality in which God is attached to the world and the self-sufficient modality in which "God is independent of all creatures" (3:97).[81] Both are known through two sets of four names. The creative set consists of the Living, the Knowing, the Willing, and the Powerful, while the self-sufficient set consists of the Living, the Speaking, the Hearing, and the Seeing. The latter relates to the Quranic verse: "Naught is as his likeness; and he is the Hearer, the Seer" (42:11). "For if he hears his speech and sees his Essence," Ibn ʿArabī writes, "surely then, his Self-existence without being related to the world is complete."[82] Together, the two modalities reveal the following seven divine names: the Living, the Speaking, the Knowing, the Hearing, the Seeing, the Willing, and the Powerful. These are unanimously accepted by traditional Islamic schools as the seven principal divine names, from which all other names derive. In premodern theological and mystical literature, they are known as the "mothers of all names" *(ummahāt al-asmāʾ)*. For Sufis, the seven principal names constitute the fundamental order of the divine presence.

Triplicity is inscribed in the seven principal names as the divine pattern of formation that complements quadrature in the creative process. Here triplicity corresponds to the three dimensions of length, width, and depth, which Ibn ʿArabī considers as embodying the productive triplicity of the creative command "Be!" The triplicity of the dimensions is revealed numerically through the seven principal names together with the *Essence*, adding up to eight, the first cubic number, being the minimum number of points required for the production of bodies in space. Explaining the structure of this pattern, Ibn ʿArabī writes:

> The ultimate aim of synthesis is the body. The body is eight points only, and what is known from the Real is just the Essence and the seven attributes. Neither are these God himself nor other than himself; likewise, neither is the body other than the points, nor are the points other than the body, nor are the points the body itself. We say that the least of bodies is eight points because the original line arises from two points up; the original plane arises from two lines up, hence the plane arises from four points; and the original body arises from two planes up, hence the body arises from eight points. The body acquired the name of the length from the line, the name of the

> width from the plane, and the name of the depth from the synthesis of two planes. Thus the body is established upon triplicity, just as the structure of syllogistic reasoning is established on triplicity,[83] and as the origin of being, the Real, becomes manifest by bringing into existence through three Realities: his Essence *(huwiyyatuhu),* his Turning *(tawajuhuhu)*, and his Uttering *(qawluhu).*[84]

The seven principal divine names also determine the six directions—up and down, left and right, front and back—further qualifying the bodies already formed on the triplicity of dimensions. Of these seven names, Ibn ʿArabī says, only six relate to the possible being, while the seventh, the Living, does not. Insofar as the Living is the condition for the existence of the other six names, it is the point where they all meet. The other six divine names—the Speaking, the Knowing, the Hearing, the Seeing, the Willing, and the Powerful—determine the six directions, while the living Essence marks the point where all coincide.[85] Thus viewed, the process of universal manifestation reveals triplicity and quadrature as complementary divine paradigms: one crystallizes the designative aspects of the creative act and the other crystallizes the productive aspects. In this process quadrature underlies the pattern of proliferation and deployment, whereas triplicity underlies the pattern of synthesis and formation.

The Human Presence

"I am the Truth," cried al-Ḥallāj, a fatal cry that was said to have led to his prosecution and tragic execution.[86] Yet the great Sufi martyr was only stating boldly what had later become the object of a sophisticated Sufi theory. That man is the mirror image of God became a central theme in Sufi thought, while *tasting* the divine reality humans embody became the prime object of the mystical experience. In the *Fuṣūṣ* Ibn ʿArabī wrote: "I praise him and he praises me, I worship him and he worships me."[87] Earlier, I referred to al-Ḥamawī's metaphor of God's Eye that never sleeps. In Arabic *insān*, "humankind," is also translated as "pupil." The term *insān* is taken to designate humankind, Ibn ʿArabī explains, because man is to God what the pupil is to the eye, the instrument of *seeing*. So if God is the light whereby the Eye sees, man is the instrument of "vision" *(baṣar)* that makes "seeing" possible. Man is *insān* because God "sees" his creatures through him, and it is the comprehensiveness of his reality that makes such vision possible.[88]

The Epitome of Creation

In al-Hindī's multiple states of Being, the first state of determination is also designated as the Muḥammadan Reality *(al-ḥaqīqa al-muḥammadiyya)*, whereas the second state is designated as the Human Reality *(al-ḥaqīqa al-insāniyya)*. In the same way these two states of Being are taken to constitute the divine presence, they are taken to constitute the human presence *(al-ḥaḍra al-insāniyya)*.[89] The human presence is but the other side of the divine presence. The logic of this hinges on the religious concept, which is not peculiar to Islam, that God created man in his image. This being so, man becomes the ultimate aim of the creation and the first thing God conceived. As God's conception of man's creation coincides with his self-manifestation, the human presence goes through the same stages of determination already discussed. These two stages can be thought of as identifying the idea of man in its ideal and particular determination. For Muslims Muḥammad is the ideal model of humankind, just as Christ is for Christians, whereas Adam is the first incarnation of this perfect model. This is what the Muḥammadan and Human Realities represent in the constitution of the human presence.

In the *Fuṣūṣ* Ibn ʿArabī explains, using numerical symbolism, the structural similarity and concordance between the divine and human presences. When the undetermined unity subjected itself to the process of determination, triplicity was the first order it revealed: unity, multiplicity, and affinity. Thus 3 was the first comprehensible form unity took. Three is also the first odd *(fard)* number, since 1 is not a number but the origin of numbers. *Fard* also means "single" and "individual," of which *farīd* means "unique." Ibn ʿArabī plays on the semantics of *fard* to show how the human presence mirrors that of the divine. In being the first object of divine knowledge, man becomes synonymous with the universe, and the articulation of the concept of 'man' in the divine mind coincides with the manifestation of the names and attributes. Thus the idea of Muḥammad, in being the model in the likeness of which man was to be fashioned—according to the *ḥadīth*: "I was a prophet when Adam was still between water and clay"—equates the divine knowledge in the indistinctive state. This is the wisdom of Muḥammad's prophethood, as Ibn ʿArabī puts it, revealing the triplicity of the state of Being with which he is identified.[90]

In his exposition of Ibn ʿArabī's *Fuṣūṣ*, al-Qāshānī explains that the notion of the Muḥammadan Reality designates the first self-determination with which the primordial Essence qualified itself. This included the determination of the hierarchy of genera, species, kinds, and individuals, the entirety of which Muḥammad comprised in his constitution. In this sense, Muḥammad was unique. Above his reality there was only the ineffable transcendent Essence.[91] The Muḥammadan Reality was particularized through the idea of Adam, with the creation of whom the human presence was realized. Adam was at once the "lens" through which God viewed all beings and the "mirror" in which he viewed his own Being. The *otherness* of the reflected image externalized divinity in a unique way, revealing the realities of the names in an embodied form.[92] If the divine presence is God revealing his names and attributes, then the human presence is the incarnation of these names and attributes in the human form. In this sense, the human presence becomes the outer face of divinity, while the divine presence becomes the inner face of humanity.[93] Al-Qāshānī explains the sense in which man is seen to epitomize the realities of the creation. Being last in the creative process, man was as it were the conclusive act, summarizing all the ontological degrees that unfolded in the process of self-determination and refocusing the colors of the ontological spectrum.[94]

In brief, the Sufi concept of the 'human presence' is based on three principles. First, man, as an idea, was the first to be conceived by God in the creative process; second, man, as an embodied form, was the last creature to be brought into existence; and third, man, in both the ideal and embodied form, constitutes the comprehensive epitome of all manifest states of Being and the sum total of all divine and cosmic realities.

Universal Man

The Sufi notion of the human presence is synonymous with that of the Universal Man *(al-insān al-kāmil/al-kullī)*. According to al-Hindī, Universal Man is the state of Being that can be attained through an ascension *(ʿurūj)*, whereby one retraces the process of manifestation back to its original source. Such ascension causes all states of Being with the expanded knowledge they entail to unfold within the individual self, resulting in transcending the limits of individuality and recognizing the universality of one's presence.[95] The notion of the Universal Man is central to Ibn ʿArabī's writings, which he articulates into a highly sophisticated theory in most of his texts, and specifically in *Inshāʾ al-Dawāʾir*, *al-Tadbīrāt al-Ilāhiyya*, *Futūḥāt*, and *Fuṣūṣ*. It is al-Jīlī, however, who is well known for his comprehensive book on Universal Man.[96] Introducing the book, al-Jīlī says that "since it was the Real *(al-ḥaqq)* who is sought in the writing of this book, it is mandatory that we deal in it with the Real-most transcendent, in regard to his names first, because they directly point to him, then in regard to his attributes because of the diversity of essential perfections they express, and because they are the first manifest forms of his revealment."[97] In sixty-three chapters, al-Jīlī goes through a wide range of metaphysical, cosmological, and eschatological themes, unfolding the dimensions of the Universal Man. Far from being confined to the domain of humanity, the concept encompasses all the realities of Being—spiritual, imaginal, and corporeal—as well as the cosmic structure in its current complexity and its reformation in the hereafter. "The Universal Man," al-Jīlī writes, "is the pole around whom revolves the spheres of Being from its beginning to its end."[98]

The Two Exemplars

The human presence, Ibn ʿArabī explains, comprises two "exemplars" *(nuskhatān)*: outward and inward. The former, temporally produced, is homologous to the whole world; the latter, eternally conceived, is homologous to the divine presence.[99] Through this dual structure man becomes at once the most universal entity and the most effective mediator between God and the world. The duality of the creator-creature is rendered interactive through the agency of man's two exemplars, since no other worldly being admits the quality of divinity, nor can the divine admit the quality of worldliness, *al-ʿubūdiyya*. Ibn ʿArabī writes:

> Man alone possesses two perfect relationships, by one he enters into the divine presence, and by the other he enters into the cosmic presence *(al-ḥaḍra al-kayāniyya)*. So he is called a "slave" with regard to his being an obligated creature that was not and then became, just like the world, and he is called "lord" with regard to his being a vicegerent, to his form, and to his being created in the best stature. Thus, he is, as it were, a mediator between the world and the Real, bringing together the created and the creator. He is the dividing line between the divine and the cosmic presences, as the dividing line between the shadow and sunlight. This is his reality: he has the perfection of both eternity and newness.[100]

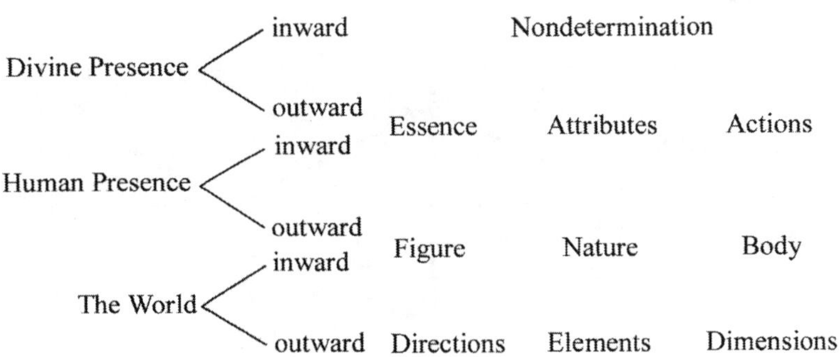

Fig. 2.9 The human presence mediating between God and the world.

Ibn ʿArabī articulates his two-exemplar concept by detailing the way in which the ternary and quaternary patterns of the divine presence are reflected in man's constitution. Here he identifies man by three essential components: nature *(ṭabīʿa)*, body *(jism)*, and figure *(shakl)*. The state of nature embodies the generative pattern of quadrature; whereas the states of body and figure crystallize the formative pattern of triplicity. The three-fold structure—nature, body, and figure—constitutes the manifest exemplar, the inner face of which corresponds to the three-fold structure of divinity—Essence, attributes, and actions. Ibn ʿArabī explains:

> In his essence, this individual man corresponds to the divine presence. God created him, in respect of his figure and organs, with six directions. These were made manifest through him because he is to the world as the point is to the circumference . . . God also created him, in respect of his nature and the form of his body, from four, so he has quadrature according to his nature, being the sum of the four elements *(arkān)*. And he structured his body *(jism)* as to have three dimensions, length, width, and depth. Thus he resembles the divine presence in regard to its Essence, attributes, and actions. These are three states: the state of his figure, which is none other than his directions, the state of his nature, and the state of his body.[101]

The two-exemplar concept allows us to see the agency of the human presence appearing in different modalities. The outward modality becomes identical

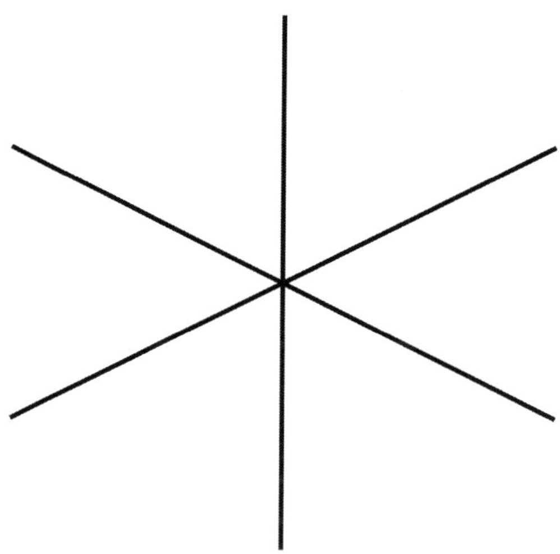

Fig. 2.10 The three-dimensional cross as a symbol of the human presence.

with the world; whereas its inward modality becomes identical with divinity. In bringing God and the world together the human presence itself tends to dissolve, just as the dividing line between shadow and sunlight that exists only through the existence of the neighboring domains.

The three states that constitute the outward exemplar of the human presence may be synthesized in the form of the three-dimensional cross, its symbol par excellence. The four arms of the horizontal cross mark the quadrature of man's nature; the three axes express the three-dimensionality of his body; and the six arms of the cross projecting from the center graphs the directionality of his unique figure. All are reconciled in the central point, which represents man's centrality in the world.[102]

Man's Nature

When God breathed his Spirit into Adam, Ibn ʿArabī says, the profusion of the Breath generated the quadrature of his nature through the four humours: yellow bile, black bile, blood, and phlegm. These four natures of the human biological structure derive from the four principal elements *(arkān)*: Fire, Earth, Air, Water. The yellow bile came from Fire, the black bile from Earth, blood from Air, and phlegm from Water. In addition, God provided man with four natural forces—attractive, fixative, digestive, and repulsive—to enable the functioning of these natures. This quaternary pattern reflects the divine creative quadrature—Life, Knowledge, Will, and Power—the model for all created quadratures.[103]

These four natures, as the four principal elements, are referred to as "*arkān*," plural of *rukn*, a "corner" or "corner pillar," giving the imagery of a quadrangular "structure." According to Ibn ʿArabī, this is the primary "structure" of being: "God established being upon quadrature and made it for himself as a house standing upon four *arkān*, for he is the first, the last, the outward, and the inward."[104] These four attributes formed the primary quadrature that necessitated the establishment of the "house" of being upon four corner pillars, within the structure of which the world of spirits and the world of bodies were manifested. From the world of divinity this structure prompted the creative attributes of Knowledge, Will, Power, and Utterance, which generated the world of spirits, which is beyond nature, as well as the natural world. The manifestation of the natural quadrature of heat, cold, moistness, and dryness followed and was employed in the generation of the world of bodies, dense and subtle. Before forming the bodies, however, God laid out the spiritual world of "writing" *(tadwīn)* and "inscription" *(tasṭīr)* that produced the original blueprint, al-Ghazālī's exemplar. This included the quadrature of the Intellect, the Soul, Nature, and Matter, through the agency of which the four elements *(arkān)* of Fire, Air, Water, and Earth were generated. It is in this divine and cosmic hierarchy that the four humors of the animal body and the four functioning forces were eventually created.[105]

As a whole constituted from the four natures, man reflects the primary divine quadrature of the first and the last, the outward and the inward in different ways.[106] With regard to God, he is the inward; with regard to the world, he is the outward; with regard to God's intention *(al-qaṣd)* in the creation, he is the first; and with regard to his existential formation *(al-nash')*, he is the last. Thus man is *first* in intention, *last* in existence, *outward* in form, and *inward* in spirit.[107] Holistically, "he is to the world as the point is to the circumference."

Man's Body

Regarding the spatial formation of an animal body, Ibn ʿArabī quotes a curious *ḥadīth* in which the Prophet is reported to have said that "the formation will be established upon the sacrum *(al-nash'a taqūm ʿalā ʿajb al-dhanab)*."[108] *Al-nash'a* is a specific Quranic term that designates at once the creation and the formation of the world—its structure, spatio-temporal conditions, and sensible forms.[109] The Quran refers to the "first creation" (*al-nash'a al-'ūlā,* 56:63), meaning the world in its current formation, and to the "other creation" (*al-nash'a al-'ukhrā,* 53:47), meaning the world as it will be *re-formed* in the hereafter. The expression *ʿajb al-dhanab* refers, according to the renowned eighth-century lexicographer and grammarian al-Farāhīdī, to the lowest point of the spine wherefrom the animal's tail originates.[110] In humans this is known as the sacrum, the triangular-shaped bone wedged between the fifth lumbar vertebra and the coccyx (tailbone), consisting of the five sacral vertebrae fused together. The sacrum is the heaviest bone of the pelvis and the center of gravity of the skeletal structure. The term comes from Latin, *os sacrum*, meaning "sacred bone," which points to its significance in medieval Europe, when it was known as the *resurrection bone*, from which a person will be reborn in the hereafter.[111] The Islamic tradition seems to have preserved this Christian conception.

The Quranic term *al-nash'a* and its link to the sacrum deserve some attention. Etymologically, it derives from the trilateral root *n.sh.'*, "to grow," "to be alive," of which *ansha'a* means "to create," "to invent," "to produce," and "to compose." The Quran says: "He *brought you forth (ansha'akum)* from the earth" (11:61). *Ansha'a* also means "to begin," "to start," and "to commence doing something." Hence the verse, "He it is who *produces (ansha'a)* gardens trellised and untrellised" (6:142) means that "he invented them and commenced their creation."[112] The term *ansha'a* has many applications in poetry and architecture. In poetry, *ansha'a* means "to compose a poem" and "to commence reciting it"; in architecture, it means "to commence setting up a structure." Therefore, *inshā'* is associated with building a structure.

The above *ḥadīth*, which many religious authorities quote, reveals a concern with the spatiality of formation: how the human body is structured and in what form it will be reconstructed in the other world.[113] Seen as the only non-perishable *(lā yablā)* component of the body, the sacrum provides the element of continuity between the two creations. Ibn ʿArabī interprets the *ḥadīth* as concerning the spatial structure of the body. He explains that the sacrum represents the *center* whence the body springs forth and upon which it is symmetrically established. It is the focal point of growth, which occurs through three centrifugal movements: downwards, upwards, and outwards. In humans, the downward movement unfolds the lower part of the body, from the sacrum to the feet; the upward movement unfolds the upper part of the body, from the sacrum to the head; and the outward movement unfolds the body in the four directions of right, left, front, and back.[114] Thus the state of man's body refers to his spatial structure in the form of the three-dimensional cross, the divine pattern of formation.

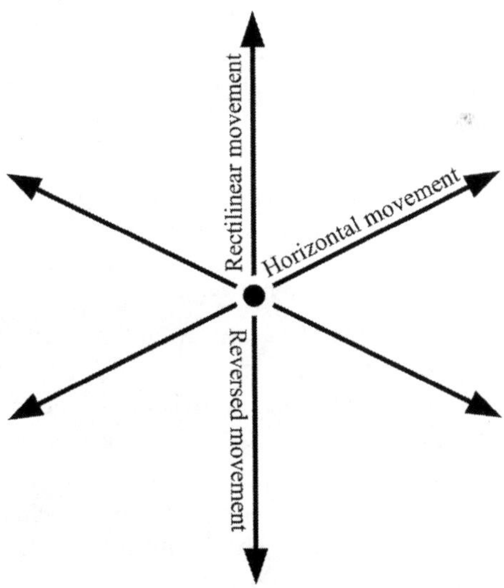

Fig. 2.11 The three movements of spatial unfolding.

Premodern Muslim physicians did not seem to have shared this view. Ibn Sīnā, for instance, describes in detail how the human body originates from the heart, the first organ of an embryo to develop in the mother's womb.[115] From the spatio-comogonic perspective, however, the sacrum seems to have been considered the center of the body. The proportional system of the human body developed by the Ikhwān in their epistle on music adds support to this idea.[116] A well-developed body, free of any kind of deformation, they posit, has a definite proportion based on its hand-span. The height from the feet to the top of the head should measure eight hand-spans, which equals the distance between the tips of the fingers when the arms are opened wide in opposite directions as a bird opening its wings.[117] This position defines a square, the center of which is a point that lies at the top of the thigh, which the Ikhwān consider to be the midpoint of the body. If the center of the square thus defined does not coincide with the sacrum, since the Ikhwān do not refer to it, it is nonetheless associated with a term that connotes the concept of 'centrality.' The Ikhwān refer to the top of the thigh bone where it joins the hip, the level on which falls the midpoint of the body, by the term *al-ḥuqq*, from *ḥaqqi*, "the middle of a thing." This term derives from the trilateral root *ḥ.q.q.*, "true" or "real," from which comes God's name *al-ḥaqq*, "Truth" or "Real." This term correlates the notion of centrality with that of reality and permanence, resonating with the *ḥadīth* of the sacrum that describes it as the only component of the human body that does not decompose *(lā yablā)*.

The state of the body of the human presence exemplifies the way in which natural bodies expand in space from their source, the point-center. Expansion occurs though physical movements of growth, which reflect the intelligible movements of manifestation, whereby the divine Essence disengages itself from primordial stillness and "moves" into the world of existence. Ibn ʿArabī describes this creative movement as the "movement of love," that is, God's passion to be known, without which the world would not have been manifested.[118] Al-Qāshānī expounds on the notion of intelligible movements *(al-ḥarakāt al-maʿqūla)*, explaining the way in which they mediate cosmic existence *(al-wujūd al-kawnī)*. Just as the sensible movements of upward, downward, and horizontal, he says, the intelligible movements designate three conceptual orientations. First is the *reversed* movement of "productive creation" *(al-takwīn)*: it is God's turning downward in order to bring the lower world into existence. Second is the *rectilinear* movement of "innovative creation" *(al-ibdāʿ)*: it is his turning upward in order to bring the worlds of the divine names and attributes as well as the worlds of spirits and souls into existence. Third is the *horizontal* movement of unfolding, turning toward the heavenly bodies, which mediate between the other two from horizon to horizon.[119] The intelligible movements of cosmic creation provide the model according to which the human body unfolds from *ʿajb al-dhanab*.

The three sensible movements of spatial unfolding are also the movements of growth tendencies. The Sufis differentiate the four kingdoms—human, animal, plant, and mineral—according to the most expressive movement in their growth tendencies. The tendency of humans is to grow upward, of the animal to grow horizontally, of plants to grow downward (their nutritive organ being the root), and of minerals not to grow, to be still. Thus humans are distinguished by their upward, ascending spatial expansion through the "rectilinear movement" *(al-ḥaraka al-mustaqīma)*. The animal is distinguished by its horizontal spatial expansion through the "horizontal movement" *(al-ḥaraka al-ufuqiyya)*. And the plant is distinguished by its downward spatial expansion through the "reversed movement" *(al-ḥaraka al-mankūsa)*. The synthesis of these three movements, together with the stillness of the mineral, reveals the three-dimensional cross as the pattern of spatial unfolding.

Along this common view, Ibn ʿArabī offers another interpretation of growth tendencies and spatial formation of natural bodies, one that is closer to the *ḥadīth* of the sacrum and al-Qāshānī's intelligible movements. He says that plants embody the reality of Growth *(numuww);* animals embody the realities of Growth and Sensation *(al-ḥis)*; and humans embody the realities of Growth, Sensation, and Reason *(al-nuṭq)*. Thus all embody the reality of Growth; however, since plants cannot intrinsically move except by way of their growth tendency, they are considered representative of the movement of growth in all natural bodies. This means that growth movement occurs in an animal body insofar as it is a plant, for other movements relate to other realities, namely, Sense and Reason.[120] Accordingly, no distinction between movements of growth as such can be made, because the body of a plant grows from the seed in the upward, downward, and outward directions exactly in the same way as an animal body grows from the sacrum. All movements of growth can thus be referred to as "rectilinear." The "reversed" movement, then, becomes the forcible movement *(al-ḥaraka al-qasriyya)*, which is contrary to the movement intrinsic to a natural object according to the law of nature, as, for example, a stone thrown in the air moving upward while its natural movement by gravity is downward. In this view, growth that leads to the formation of bodies in space is considered to occur through the following movements: first, movement from the center *(ḥaraka min al-wasat)*, that is, the simultaneous, centrifugal movement or spatial expansion from the origin in all directions; second, movement to the center *(ḥaraka ilā al-wasaṭ)*, that is, the simultaneous, centripetal movement of divine sustenance that determines the extent of growth in each direction; and third, movement within the center *(ḥaraka fī al-wasaṭ)*, that is, the essential enlivening movement whereby the essence of the origin subsists.[121]

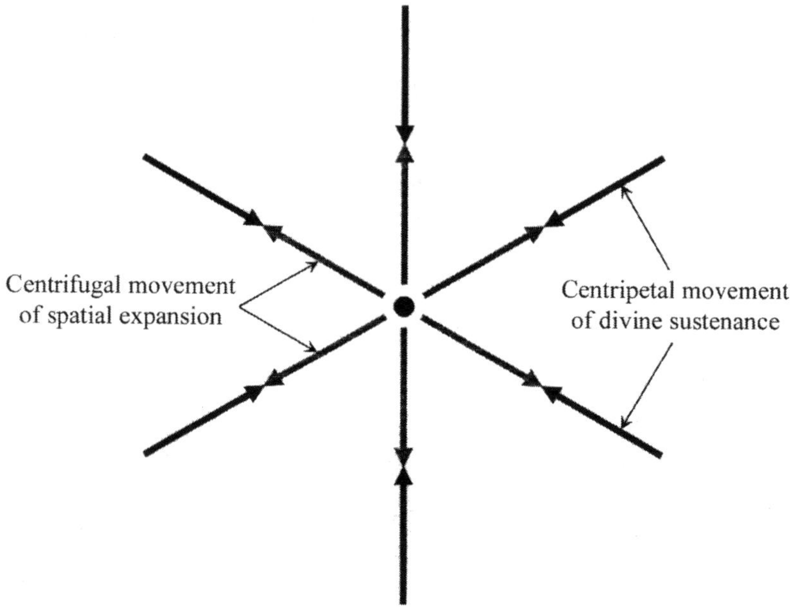

Fig. 2.12 The complementary movements of spatial expansion according to Ibn ʿArabī.

Man's Figure

In the well-known story of Adam's creation, the Quran tells how God informed the angels and the jinn that he was about to set on earth a viceroy before whom they were to prostrate themselves in deference to his superiority. Somewhat baffled, they all did except Iblīs who, taking pride in his fiery nature, unrepentantly refused. Having been expelled from paradise because of his rebellious, disobedient attitude, Iblīs revealed his sinister intentions: "Now, because you have sent me astray, verily I shall lurk in ambush for them on your right path. Then I shall come upon them from before them and from behind them and from their right hands and from their left hands, and you will not find most of them grateful" (7:16–17). The reference to man's directions in this dialogue is rather curious. Why is man identified by his directions? Why only the four horizontal directions? What about the above and the below? What does it mean for Satan to approach man through these directions? Why is Satan's attack spatially referenced? These questions lead us to consider man's figure. The three-dimensional cross identifies at once the quantitative and qualitative aspects of the human presence. The state of man's body *(jism)* has led us to consider the quantitative aspect, the *dimension*, now the state of man's figure *(shakl)* leads us to consider the qualitative aspect, the *direction*.

In themselves, directions in space are indifferent. The sphere as a spatial expression of all possibility contains all possible directions, which are determined by extensions from the center to the surface. Indefinite in number, all directions are also equal in significance. The human figure, however, in being structured upon six qualitatively different directions, qualifies space by rendering its directions significantly different. Commenting on the Quranic verse, "The originator of the heavens and the earth" (2:117), Ibn ʿArabī says that heaven is what ascends and earth is what descends, and man is the one who distinguishes between what is above and below because he is the one with the directions.[122] The spatial formation of man distinguishes six main directions: front and back, right and left, up and down. This differentiation is based on bodily attributes and functions. Front is the direction of vision toward which man naturally moves; back is the direction of the unknown, of vulnerability; right is the direction of strength, being naturally the stronger side; left is the direction of weakness, being naturally the weaker side; up is the direction of man's head, pointing heavenward; and down is the direction of his feet, pointing earthward.[123]

Identifying man by his directions in the above verse assumes certain links between directions and one's virtues and beliefs. So Satan's approach from the front is understood as making people indulge in the pleasures of the world they see, whereas his approach from the back is understood as making people doubt the reality of the world they do not see, the hereafter. His approach from the right is the corruption he may bring to the soul through its good virtues and from the left through its bad virtues. The above and down were inaccessible by him, for they designate the vertical channel of God's direct communication and mercy.[124]

Ibn ʿArabī takes this understanding further. First, he sees in Satan's spatial reference an indication of human superiority. Satan's reference to four directions only, he says, is an expression of the limitation of his formation and of his world. Lacking the reality of transcendence that grants him access to the vertical axis, he remains inferior to man.[125] Man's upright bodily formation generated by the rectilinear movement, as we have seen, sets him apart from all other creatures. This awareness of the uniqueness of verticality has a spiritual significance in the mystical experience. The concept of 'verticality' is viewed to be a spatial expression of the Muḥammadan Reality in its eternal presence. The ninth-century Sufi Sahl al-Tusturī (d. 896) speaks of the differentiation of the "light of Muḥammad" *(nūr Muḥammad)* from the divine light in spatial terms. When God intended to create Muḥammad, he says, he projected from his own light a distinct light *(aẓhara min nūrihi nūran)*. "When it reached the veil of the Majesty *(ḥijāb al-ʿaẓama)* it bowed in prostration before God. God created from its prostration *(sajda)* a mighty column *(ʿāmūd)* like crystal glass *(zujāj)* of light that is outwardly *(ẓāhir)* and inwardly *(bāṭin)* translucent."[126] It is from this Muḥammadan light, al-Tusturī adds, that the human race originated. Adam was the first to be manifested in this way: "God created Adam from the light of Muḥammad."[127] Before this, the Muḥammad of preexistence had stood as a column before God for a million years "without body *(jism)* and form *(rasm)*."[128] "When the preexistential and temporal universe as well as the prophetic[129] and spiritual prototypes had completed the emanation of light ultimately from Muḥammad's light, Muḥammad was shaped in a body *(jasad)*, in his terrestrial form, from the clay of Adam *(ṭīn Adam)*. This clay of Adam in turn had been formed from the column of light in which Muḥammad had served his Lord in preexistence."[130] This column of light, which is "as thick as the seven heavens,"[131] is the archetype man's upright posture embodies.

The notion of the 'column of light' emphasizes the significance of the vertical axis and its exclusive association with man. The state of the human figure is concerned with the meaning of this spatial uniqueness. The Quran speaks explicitly of the human superiority over all creatures and especially over the other two rational creatures, the jinn and the angels. As a living, rational creature with a sensible body, man gathers together in his formation the qualities of two kinds of creatures: creatures endowed with the rational faculty but lacking a sensible body and those with a sensible body but lacking the rational faculty. The former includes the jinn and the angels, and the latter includes the animals and the plants. None of these creatures shares with man one single quality: his verticality. Animals are characterized by their horizontality, plants by their downwardness, and humans alone by their upwardness. Humans share the qualities of horizontality and downwardness with animals and plants, whereas their verticality renders them distinct. Lacking the rational faculty, animals and plants cannot become consciously aware of the significance of their spatial structure, even if it were the same as that of humans. Humans' superior spatial structure gains further significance when coupled with their rationality.

In Satan's attack on man through the four horizontal directions, Ibn ʿArabī sees a metaphor for depraving and corrupting human psychic characteristics that are associated with these directions. The front is the direction of vision, of the known, so it is associated with confidence and certainty as humans are in command of what happens in front of them. Satan attacks them from this direction by making them skeptical and uncertain, so they may doubt the oneness of God and become polytheists *(mushrik)*. The back is the direction of the unseen, of the unknown, so it is associated with ignorance and fear. Satan attacks people from this direction by exploiting their ignorance and making them disbelievers or making them believe only in the incomparability of God (*muʿaṭṭil*, one who refuses the analogical relationships between God and the created world). The right is the direction of strength, so it is associated with power and authority. Satan attacks people from this direction by weakening them, and by exploiting human authority to make them arrogant *(mutakabbir)*. The left is the direction of weakness, so it is associated with pretence and dependence. Satan attacks people from this direction by exploiting their pretentiousness to make them hypocrites *(munāfiq)*.[132]

The Satanic attack finds support in human earthly nature and sensuous desires. Therefore, Ibn ʿArabī says, people are ordered to fight him from these directions, which should be fortified according to what the law *(al-sharʿ)* has ordered them to fortify them with, so Satan would not find a way to approach them.[133] The fortification of these directions takes on cosmic dimensions in the Sufi teaching, wherein four spiritual masters (*awtād*, "pegs" or "pillars") are identified with the four directions—east, west, north, and south. By these four "pillars," Ibn ʿArabī explains, God preserves the four cardinal directions, one pillar for every direction. And by these four pillars together with the "pole," *al-quṭb,* the greatest master who represents the cosmic axis, God preserves the existence of the world.[134] Satan has no access to the upward and downward directions because of his limitation. The exclusive verticality people have constitutes their transcendental dimension. It is the dimension that enables them to transcend the horizontality of their animality, to communicate with heaven, and to receive divine pure inspirations free from satanic contaminations. The above, Ibn ʿArabī says, "is the direction that leads toward the spirit, from which comes truthful inspirations and angelic revelations, and from which knowledge and spiritual realities emanate."[135]

In the context of Sufi teachings, verticality is a spatial expression of human uniqueness, while the six directions comprise an expression of the comprehensiveness of the human reality. By giving meanings to the directions in space, such teaching engenders a particular spatial sensibility based on an awareness of the psycho-religious significance of directions. The three dimensions and six directions are the spatial conditions that were seen to govern the entire natural world. Along with human nature, one's body and figure exemplify these conditions, providing a constant reminder of the foundational order of spatial existence.

The Presence of the Word

In al-Ghazālī's analogy, God did not *draw* the blueprint of the world; he *wrote* it. Although drawing might appear to us as more universal than writing and an image more expressive than a word, *inscribing*, in a sense, conflates drawing and writing, the image and the word. In Ibn ʿArabī's cosmogony, as we have seen, it is the enunciation of *kun* that brought the world into existence: in the beginning was the word. The transcendent Essence that affirms its unity through the numerical one and the geometrical point also reveals itself as the word. The primordial word, the divine logos, in its uttered and inscribed modes, is the primary means by which the world was actualized. This archaic view has continued to thrive within the Islamic tradition taking on new dimensions.

In Islam the divine word was seen as being incarnated in the sacred text of the Quran; hence it is only natural to speak of the presence of the word. Schuon draws our attention to the ubiquity and profound influence of the Quranic text when he says: "The verses of the Quran are not only utterances which transmit thoughts; they are also, in a sense, beings, powers, talismans. The soul of the Muslim is as it were woven out of sacred formulae; in these he works, in these he rests, in these he lives, in these he dies."[136] This certainly resonates with the Sufi perspective, which views the language of the Islamic revelation, Arabic, in both its written and oral forms, as an embodiment of the primordial word, a materialization of the creative enunciation. The underlying patterns of divine realities, discussed above, find immediate expressions in this domain. The concept of the Unity of Being and all the states it contains, the divine and the human presences, all find immediate correspondences in various aspects of the Arabic language.[137] In the world of utterance and inscription, beings take on, so to speak, a linguistic guise. They become, as Ibn ʿArabī puts it, "letters inscribed in the spread parchment of existence wherein writing is ceaseless and endless."[138] The Sufi teachings on the symbolism of the letters have survived well into the twentieth century through figures such as the Algerian shaykh Aḥmad al-ʿAlawī (d. 1934), who wrote in the true spirit of the Sufi tradition. His tract on the symbolism of the letters reveals the profundity of Ibn ʿArabī and al-Jīlī.[139]

The World as a Book

Mir'āt al-ʿĀrifīn, a popular treatise on the meanings of the Quran's opening chapter *(al-fātiḥa)* attributed variably to al-Qūnawī, Ibn ʿArabī, and even *imām* al-Ḥussayn, opens by saying: "Praise be to God who externalized from the *nūn* (N) what he internalized in the Pen, and brought out into being by benevolence what he treasured in non-Being . . . And glory to him who . . . unrolled the parchment of the world *(al-raqq al-manshūr)* and inscribed the archetypal book *(al-kitāb al-masṭūr)* by the ink of existence, which manifests all that is latent within the speaker in the form of letters and perfect words."[140] The metaphor of the world as a book is common in premodern Islam. There are numerous Sufi treatises devoted to the science of letters, *ʿilm al-ḥurūf,* whose origin is often attributed to the fourth caliph and the Prophet's son-in-law Alī bin Abī Ṭālib. The Quranic imageries of the pen, the ink-well of *nūn*, and the divine act of writing provide the basic conceptual tools used by Sufis and other theologians in the development of their metaphorical interpretations. The concepts of the 'Pen' *(al-qalam)* and the 'Preserved Tablet' *(al-lawḥ al-maḥfūẓ),* the analogy of the trees as pens and the seas as ink, of the word as a tree, and so on, along with the prophetic traditions that corroborate these Quranic ideas all form the foundation of alphabetical symbolism in Islam. There are also the fourteen mysterious disjointed letters that appear at the beginning of several Quranic chapters. These received great attention in premodern Islam and contributed significantly to the science of alphabetical symbolism. Forming exactly half of the twenty-eight letters of the Arabic alphabet, these fourteen disjointed letters are seen as representing the spiritual dimension of the alphabet, corresponding to the world of spirits. They are called the "luminous letters" *(ḥurūf nūrāniyya)*, in contrast to the other fourteen that are taken to represent the corporeal dimension and are, therefore, called the "tenebrous letters" *(ḥurūf ẓalmāniyya)*. The science of the letters, Ibn ʿArabī explains, concerns both the "length" and the "breadth" of the world. The "length of the world" *(ṭūl al-ʿālam)* refers to the spiritual world, the world of meanings, whereas the "breadth of the world" *(ʿarḍ al-ʿālam)* refers to the physical world, the world of bodies.[141] This resonates with his interpretations of the spatiality of the human body already discussed.

In the parallels Sufis draw between the world and the Quran, letters and words acquire individual presences just as other beings do. Everything is brought forth through the creative enunciation mediated by the "Breath of the Compassionate" *(al-nafas al-raḥmānī)*, the substance of life pervading the universe. In this sense, each letter or sound becomes an entity in its own right, determining and articulating the undifferentiated "sound" of the creative enunciation.[142] Their manifestation coincides with the utterance of the creative command "Be!" *(kun)* and the exteriorization of the "cosmos" *(kawn)* in the forms of letters, words, sentences, and texts. The presence of the word thus emerges from seeing all cosmic entities as phonetic articulations manifested through the articulation of the divine Breath. *Mir'āt al-ʿĀrifīn* says:

> Every being is a letter (*ḥarf*) in a sense, a word (*kalima*) in a sense, an isolated, disjointed letter (*mufrad wa muqaṭṭaʿ*) in a sense, a composed utterance (*alfāẓ murakkaba*) in a sense, and a *sūra* in a sense. When we consider only the essence (*dhāt*) of every being without considering its aspects (*wujūh*), properties (*khawāṣ*), accidents (*ʿawāriḍ*), and concomitants (*lawāzim*), as dissociated

> from the whole, we call it, with reference to this dissociation, a "letter." And when we consider its aspects, properties, accidents, and concomitants in association with the essence, we call it, with reference to its association with the whole, a "word." And with regard to the abstraction of every being from the additions and relations, and to the distinction from one another, they are called "isolated, disjointed letters."[143] And with regard to the nonabstraction of beings from the additions and relations, and to the nondistinction from one another, they are called "composed utterances." And with regard to the distinction of the universal states of Being from one another, and to every being falling under one state, they are called "chapters" (sura).[144]

The attribute of knowledge, as already discussed, is the first determination of the divine Essence. In the context of alphabetical symbolism, manifestation becomes the "book" that contains the divine knowledge. Here Sufis articulate two concepts concerning the detailed and summarized versions of the "book." "Know O well-supported son that the world is two-fold, the world of command and the world of creation, and that each is a book from God's many books, and that each has an opening, and that all that is detailed in the book is summed up in the opening. So with regard to summing up what is detailed in the book, the opening is called the 'mother of the book' *(umm al-kitāb),* and with regard to unpacking what is summed up in it, this state of detailing is called the 'clear book' *(al-kitāb al-mubīn).*"[145]

Both concepts derive from the Quran, which is referred to as the "clear book" and its opening chapter (*al-fātiḥa,* "that which opens") as the "mother of the book." A *ḥadīth* takes this process of miniaturization further to the point of the first letter.[146] The opening chapter comprises seven verses that are seen as corresponding to the seven principal divine names, which are called the "mothers of the names." Just as these seven names contain all the divine names, so likewise *al-fātiḥa* contains in synoptic form all the truths revealed in the book.[147] Understood as signifying the potential and actual modes of being, the Sufis have applied both concepts at various levels of existence. Consistently, the "mother of the book" refers to the maternal source wherein all is potentially latent, whereas the "clear book" refers to the projected state where the undifferentiated totality is revealed in differentiated forms. At the divine level, the Essence, in that all divine realities are latent within it, is designated as the "mother of the book," whereas God's knowledge of himself, which reveals these realities in the form of the names, is designated as the "clear book." In the world of archetypes, the Pen, in that all cosmic realities are latent within it, is designated as the "mother of the book," whereas the Preserved Tablet, which reveals these realities as cosmic forms, is designated as the "clear book." In the world of nature, the Throne *(al-ʿarsh),* in that all the realities of the physical world are latent within it, is designated as the "mother of the book," whereas the Footstool *(al-kursī)*, which reveals these realities in the forms of the heavens and the earth, is designated as the "clear book."[148]

The twenty-eight letters of the Arabic alphabet are viewed to correspond to the "human formation" *(al-nash'a al-insāniyya)* in both its bodily and spiritual constitution. "Jawāhir al-Sirr al-Munīr," a Sufi treatise on the symbolism of the letters attributed to Ibn Sabʿīn, shows in a diagrammatic way how every letter corresponds to one part of the human body. While the outward forms of the letters correspond to the human body, the "Jawāhir" says, their inner meanings correspond to the human spirit. The twenty-eight letters of the Arabic alphabet are God's secrets in the world. "They are formed in the image of a human figure, as a person standing upright, whose creation is perfect, that is, composed of two parts: spirit and body."[149]

The "Jawāhir" divides the letters into four kinds: intellectual *(fikriyya)*, uttered *(lafẓiyya)*, written *(raqamiyya)*, and numerical *(ʿadadiyya)*. Two of these

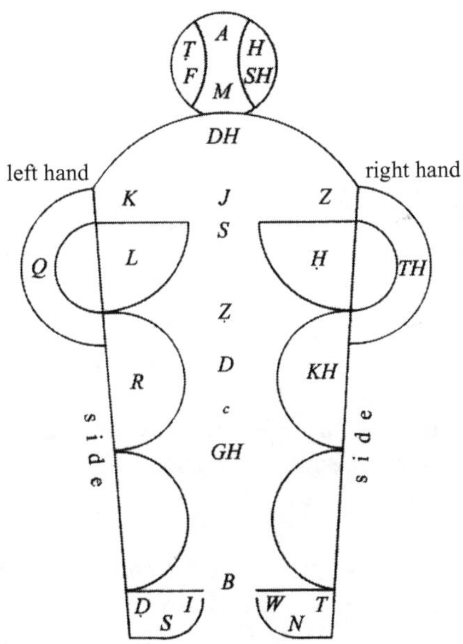

Fig. 2.13 The correlation of the Arabic alphabet and the human body ("Jawāhir," MS. 7127).

Fire hot and dry	Earth cold and dry	Air hot and moist	Water cold and moist
a	*b*	*j*	*d*
h	*w*	*z*	*ḥ*
ṭ	*y*	*k*	*l*
m	*n*	*ṣ*	ʿ
f	*ḍ*	*q*	*r*
s	*t*	*th*	*kh*
dh	*ẓ*	*gh*	*sh*

Fig. 2.14 The natural qualities of the Arabic alphabet ("Jawāhir," MS. 7127).

kinds, the uttered and the written, are considered to be manifest, and the other two, the intellectual and the numerical, to be hidden. Thus, the latter are seen to be "in the state of the spirit," and the former "in the state of the body."[150] In addition to expressing the bodily and spiritual dimension of the human presence, the Arabic letters are also viewed to have different natures, whereby they embody the creative quadrature and correspond to the four *arkān*—Fire, Air, Water, and Earth. For example, the letter *alif* (A) is considered to be hot and dry, corresponding to Fire; *bā'* (B) is cold and dry, corresponding to Earth; *jīm* (J) is hot and moist, corresponding to Air; and *dāl* (D) is cold and moist, corresponding to Water. There are several systems that classify the letters according to their natural qualities.

With regard to their calligraphic forms, the letters are seen as being composed of the primary geometric forms: the point, the line, and the circle. The *alif* (A), for example, is a vertical line; the *bā'* (B) is a horizontal line with a point underneath; the *nūn* (N) is half a circle with a central point; the *lām* (L) is half a circle with a vertical line on one end; and so on. In this way the Sufis extend their geometrical symbolism to the alphabet, thereby conflating writing and drawing under the notion of *inscription* as well as relating the spatial order of the letters to that of the divine and the human presences. The Arabic letters also have numerical values that play a significant role in the Sufi interpretations. The geometrical point or the alphabetical dot *(nuqṭa)* is where it all begins. The Sufis teach that just as all existents are conceived within the primordial Essence, so likewise all letters, words, sentences, and texts are contained within the primordial "dot." And just as all beings are manifested and differentiated from the incomprehensible Being, so likewise all the letters, words, sentences, and texts are manifested and differentiated from the impenetrable point.

The first two letters of the Arabic alphabet, the *alif* (A) and the *bāʾ* (B), present in the presence of the word the traces of the universal realities and order of Being as revealed in the divine and human presences.

The Alif

The letter *alif* (A), written as a vertical stroke (|), is the first letter of the Arabic alphabet. The term *alif* derives from the root *a.l.f.*, "thousand," the verb of which, *allafa*, means "to bring together," "to attune," "to harmonize," and "to compose." The nouns *ilf* and *ilfa* mean "familiarity," "intimacy," and "harmony." Al-Jīlī relates the meaning of *alif* to the human feeling of closeness and familiarity *(ilfa).* The first letter was named "*alif*," he says, because, just like *ilfa* brings people closer together, it brings all the letters together by forming their shared inner substance.[151] Here al-Jīlī refers to the names and pronunciation of the Arabic letters, which, in one form or another, contain the *alif.* Insofar as the *alif* is a geometrical line, all letters, as geometrical shapes, can also be reduced to it. Al-ʿAlawī explains how the spatial formations of the letters are no more than a transformation of the *alif.* The *hāʾ* (H), he says, is simply a hunchback *alif,* while the *mīm* (M) is a circular *alif.*[152] The *alif* is what all the letters have in common.

Islamic mythology provides many interesting narratives on the creation of the *alif.* The "Jawāhir" says that God first created the Pen from a green emerald and the Tablet from white light and then ordered the Pen to inscribe onto the Tablet the destiny, or his knowledge, of the created world. Upon this divine order a "drop" (*nuqṭa*, "point") fell from the nib of the Pen. It overflowed inscribing a line standing upright. When God saw this he decided to make it the first letter of his exalted name *Allāh*. The *alif* thus became the origin of all the letters just as God's generosity was the source of all existents.[153] Al-ʿAlawī overlays the same narrative with a poetic imagery: "Indeed the *Alif* is none other than the Point itself which is an eye that wept or a drop that gushed forth and which in its downpour was named *Alif.*"[154]

According to Ibn ʿArabī, the *alif* has two forms: one in writing as a vertical stroke, the other in utterance as the *hamza* (hiatus). For him the *alif* is not a letter but the origin of all letters, just as 1 is not a number but the origin of all numbers. In utterance, the *alif* is the unobstructed breath emanating from the heart, whose various guttural, palatal, dental, and labial articulations manifest the letters of the alphabet. Numerically, the *alif* is number 1; geometrically, it is the line; and calligraphically, it is the diameter of the circle within which the other letters are differentiated. Accordingly, the *alif* represents the first definable form of unity that emerged from the undefinable point. Unlike all other letters, al-Jīlī says, the *alif* is "only one degree distant from the point, for two points together make an *alif*."[155] It is the "first definable appearance" of the point. By appearing in the form of the *alif* the point qualifies itself with *firstness*. Thus the *alif* stands for the state of first determination, for divinity in its first knowable state.[156]

As the first affirmation of unity, the *alif* corresponds to the Muḥammadan Reality, the "column of light," in al-Tusturī's terms. Al-Jīlī finds reference to this in the *ḥadīth* that says that the first thing God created from his Essence was the spirit of Muḥammad, and from this spirit he then created the entire world. Here the *alif* becomes a visual evidence of the first stage in the creative process. "Every letter is composed from the point, so the point is as a simple substance *(jawhar basīt)*, while the letter is as a composed body *(jism murakkab)*. The *alif*, in that every letter is shaped from it, represents the point. So in its composed form, the *alif* represents the simple substance of the point, because all the letters are shaped from it . . . And so likewise the Muḥammadan Reality from which the entire world is created."[157] The analogy between the manifestation of the world and the differentiation of the letters is a common theme in the Sufi literature. In the same way the manifestation of the divine presence was not caused by anything other than the irradiation of Essence itself and its inward love to be known, so was the manifestation of the *alif* caused by the overflowing of the point. The original *alif* was "not traced by the pen, nor was it dependent upon it, but sprung from the outward urge of the Point in its principal centre."[158] The act of overflowing brings out the *alif* without any detriment to the integrity of the flawless dot that remains transcendent in its eternal incomparability.

The **Bāʾ**

The letter *bāʾ* (B), written as a horizontal line with a point underneath it (ب), is the second letter of the Arabic alphabet. It is the first letter of the first word in the Quran, *bism*, "in the name," with *al-basmala* considered as the first verse. Two traditions frequently quoted by the Sufis form the basis of alphabetical symbolism in general and the *bāʾ* in particular. The first says: "All that is in the revealed books is in the Quran, and all that is in the Quran is in the *fātiḥa*, and all that is in the *fātiḥa* is in *bism Allāh al-raḥmān al-raḥīm*"; the other says: "All that is in *bism Allāh al-raḥmān al-raḥīm* is in the letter *bā'*, which itself is contained in the point that is beneath it."[159] The *bā'* is seen to represent the first differentiation of the *alif*. It is its first articulated form with which the *alif* uniquely appears.[160] Thus the *bā'* is taken to stand for the human presence, the Universal Man, that is, the outward mode of the divine presence. The *bā'*, the Sufis teach, is nothing but man, the "first man," who is "the spirit of being," created in the image of God, the *alif*.[161]

As a horizontal extension, the *bā'* graphs the shadow of the vertical *alif* standing before the radiating light of the point. As the shadow of the *alif*, the *bā'* carries within it a visible trace of the original source, which is the point that appears beneath it. The point of *bā'* becomes the shadow of the higher point that resides "in its hidden-treasurehood" before its first self-disclosure as an *alif*.[162] The transcendental point that lies above the *alif* descends to appear underneath the *bā'*, just as divinity images itself in the human form. The Sufis see in this a reaffirmation of the universal realities and a clear illustration that the things of the lower worlds are manifestations of the things of the higher worlds. They refer to the prophetic tradition that says: "If you lower a rope unto the nethermost earth it would light upon God," to show how the *bā'* discloses the truth that underlies all things. They also refer to the verse that says that "everything perishes but his Face," to illustrate how the alphabetical symbol of the human presence at one veils and reveals in its form the unperishable face of divinity.[163] The point beneath the *bā'* becomes the seal of divinity in the created world, a constant reminder of the origin *(aṣl)* whence everything proceeds.

Despite its veiled appearance in the *bā'* the point remains essentially distinct from the letters, in the same way that Being, despite manifesting in all other beings, remains "nothing is as his likeness" (42:2). Al-ʿAlawī writes: "The point was in its principal state of utterly impenetrable secrecy where there is neither separation nor union, neither after nor before, neither breadth nor length, and all the letters were obliterate in its hidden Essence."[164] And even though it reveals itself in the form of all the letters, the point remains above "all that is to be found in the letters by way of length and shortness and protuberance,"[165] and beyond the grasp of vision, aurality, or literacy.

The Soliloquy of the **Bā'**

A Sufi tradition says that "by the *bā'* Being manifests; and by the point the adoring is distinguished from the adored."[166] In "al-Kahf wa al-Raqīm," al-Jīlī reflects on the spatial form of the *bā'*, presenting a dialogue between the letter itself and the point that lies beneath it. The conversation reminds us of al-Nābulusī's soliloquy, while presenting the doctrine of the Unity of Being in a geometrical guise. On the one hand, al-Jīlī's soliloquy illustrates the meaning of the above Sufi tradition, and on the other, it expresses in a reflective, symbolic manner the bounding relationship between God and man, the divine and the human presence. He writes:

> The point says to the *bā'*: "O letter, I am your origin *(aṣl)* because you are composed from me. Yet, in your composite form, you are my origin, because every part of you is a point, so you are the whole *(al-kull)* while I am the part *(al-juz')*: the whole is the origin while the part is the branch *(far')*. In reality, however, I am the origin, because your composition is none other than me *('ayn)*. Do not look at my appearance behind you and say: 'This manifest thing is other than me,' for I regard you to be none other than myself and my identity *(huwiyya)*. And had I not existed in you I would not have had such a relationship to you. Until when will you turn your perception *(shahāda)* away from me, leaving me behind your back? Make your hidden mysteries be your perception and your perception be your hidden mysteries by realizing my unity with you. Without you I would not have been the point of the *bā'*, and without me you would not have been the *bā'* with a point. How many symbols have I struck for you so that you may understand my unity with you, and know that your expansion *(inbisāṭ)* in the world of the seen *('ālam al-shahāda)* and my concealment *(istitār)* in the world of the unseen *('ālam al-ghayb)* are two modalities for our same essence. No one participates in my relationship to you, nor in your relationship to me. You are not 'you,' because your name is novel compared to mine. Can you not see how the first part of you is called 'point,' the second part is called 'point,' the third part is called 'point,' and so are the rest of your parts, point by point. I am you; you have no I-ness in yourself. Rather, my identity is your I-ness whereby you are what you are. Had you, when saying in yourself I, imagined my essence, I, too, would have, when saying he, imagined my face *(wajh)*. Then, you would know that 'I' and 'he' are two expressions for one essence."
>
> The *bā'* said: "O master, it has become certain to me that you are my origin, and I have realized that the branch and the origin are the same. This is my body extended and composed; I cannot exist except within it. I am a gross body *(jism kathīf)* bound to one place only whereas you are a subtle substance *(jawhar laṭīf)* that can exist in everything. So how could I have the reality of yours? How could I be you? How would your conditions be the same as mine?"
>
> The point answered the *bā'* and said: "perceiving your corporeality and imagining my spirituality is a form and a modality of mine. And since all the various letters and words, in their entirety, are none other than me, how could there be a distance? And even though the ten cannot be regarded as the name of the sum of these five units, where, in the reality of the ten, would the differ-

ence between the five and the ten be except in the name-ness *(al-ismiyya)*? You are, with all your aspects, being a modality and a glance of mine; where would the polarity between you and me be? and how? while I am not only the origin of this dialogue between you and me but also of all of that which comes out of you and me. All of this is none other than myself: an order of a divine wisdom. So if you want to conceive of me, imagine yourself, the letters, all of them, and the words, small and large, then say point, that totality is none other than myself, and myself is none other than that totality."[167]

The Formation of the Word

Describing the creation of Adam, the Quran reports: "So, when I have formed him *(sawwaytahu)* and have breathed into him of my spirit" (15:29). With reference to this verse, Ibn ʿArabī compares the formation of man to the formation of the word, that is, to the way in which letters are joined together to form meaningful, utterable words. A brief clarification of the nature of the Arabic is necessary here. Unlike the Latin-based alphabets, the Arabic alphabet does not contain, properly speaking, letters that are designated as vowels. The twenty-eight letters of the Arabic alphabet are consonants and unutterable on their own. The letters *a*, *w*, and *y*, which are usually referred to as "long vowels," are in Arabic *ḥurūf al-ʿilla*, literally, "the letters of weakness," of "deficiency," or of "cause" (the philosophic expression *al-ʿilla al-ʾūlā* means the "prime cause"). There are countless words in Arabic in which these letters form no part; hence, they are not vowels in the literal sense of the word. Instead, there are six *ḥarakāt*, literally "motions," marked on words in the form of diacritical notations that play the role of vowels in Arabic. They are not letters, however, nor do they form part of the alphabet. In a word such as *ḌaRaBa*, "to strike," for example, the only letters that are written are those of the trilateral root *Ḍ.R.B.*; to put this same verb in a passive form *ḌuRiBa*, "is struck," changes nothing in the word's spelling. The only things that change are the unwritten "phonetic motions" *(ḥarakāt)*, without which the root is unpronounceable, meaningless, or "dead," so to speak. Utterance that causes a word to exist, to assume a presence, to be alive, is effected through the application of the *ḥarakāt*. The letter is unutterable if not mobilized by the vocalizing motions.

From this perspective, the consonant letters of the alphabet are viewed to constitute the word's lifeless body whereas vowelling to act as the animating spirit. It is from this perspective that Ibn ʿArabī compares the addition of the phonetic motions onto the letters, after their being prepared *(taswiya)* to receive these motions, to the formation of Adam as described in the above verse. Through the agency of motions, Ibn ʿArabī says, the letters are brought forth in a new formation *(nash'a)* called "word" *(kalima)*, just as any individual of us is called "man" *(insān)* only after receiving the divine spirit.[168] This process also corresponds to the initial stage of the cosmogonic process, when the world is disengaged from the stillness of the primordial chaos, the state in which the possibilities of manifestation, still virtual, are lost in the indifferentiation of its *materia*. Ibn ʿArabī writes: "God first brought the entire world into existence in the form of a well-prepared *(musawwa)*, yet lifeless, ghost. It was like an unpolished mirror. But it is a rule in the divine business to prepare no place without it being able to receive a divine spirit, an act referred to as the 'blowing of spirit into it.' This is none other than the already prepared form reaching a state of readiness *(istiʿdād)* to receive the incessant, radiating effusion *(al-fayḍ)* that has been and will always be."[169]

The phonetic system of the Arabic language forms the basis of the Sufi notion of the formation of the word. The letters of the Arabic alphabet are seen to represent, insofar as they are all consonants, a homogeneous substratum that does not yet include any qualitative or differentiated imprint. The addition of the vocalizing motions *(ḥarakāt)* to the letters symbolizes the blowing of the spirit into this homogeneous substratum, an act that disengages the letters from the stillness of their primordial consonance, bringing them forth into the audible world of sound.

In grammatical terms, the six phonetic motions are divided into two correlated sets of three. One is *ḥarakāt al-iʿrāb*, literally, "motions of expression"; the other is *ḥarakāt al-binā'*, literally, "motions of building." They are correlated in the following order:

vowel	*iʿrāb*	*binā'*	literal meaning
A	*naṣb*	*fatḥ*	"erecting," "unfolding"
U	*rafʿ*	*ḍamm*	"rasing," "embracing"
I	*khafḍ*	*kasr*	"bringing down," "breaking"

A consonant letter that is not subject to any of these phonetic motions is grammatically identified with *sukūn*, "stillness."[170] The pattern of formation constituted by the three phonetic motions—"unfolding" *(fatḥ)*, "raising" *(rafʿ)*, and "bringing down" *(khafḍ)*, together with "stillness" *(sukūn)* as the common center whence these "motions" emanate—retraces the Sufis' pattern of cosmic existence, spatial unfolding, and natural growth already discussed. Ibn ʿArabī says the *fatḥ* signifies the unfolding of existence, *rafʿ* signifies transcendence, and *khafḍ* signifies corporeality.[171] They correspond to the horizontal, rectilinear, and reversed movements respectively, revealing the three-dimensional cross, the pattern of triplicity. In the *Fihrist*, the tenth-century scholar and biographer Ibn al-Nadīm quotes the ninth-century scholar Sahl b. Hārūn as saying: "*Al-iʿrāb* is made up of three motions *(ḥarakāt)—al-rafʿ*, *al-naṣb*, and *al-khafḍ*—because the natural movements are three: (1) movement from the center, like the movement of fire; (2) movement to the center, like the movement of earth (movement caused by gravity); and (3) movement about the center, like the movement of a sphere."[172]

The act of adding these phonetic motions to the letters is called in Arabic "*tashkīl*," literally, "giving shape, morph or figure," and "forming." It derives from *shakl*, literally, "shape," "morph," and "figure." Thus the act of transforming the consonant letters into pronounceable words connotes the idea of forming or shaping, giving, as it were, sonic-audible forms to the synthesis in the same way the human body receives its spatial-visual form when brought into existence. Ibn ʿArabī says: "Such is the way the world of words and utterances is formed from the world of letters. The letters are matter for words, just as water, earth, fire, and air are matter for the formation of our bodies."[173] And just as nothing moves in the world except by the order of the immovable principle, so likewise in the world of letters, no phonetic motion may ever manifest except by the order of the principial stillness. "The promptings unto utterance," Al-ʿAlawī writes, "were set in motion according to the demands of the Point's attributes which lay hidden in its Essence."[174]

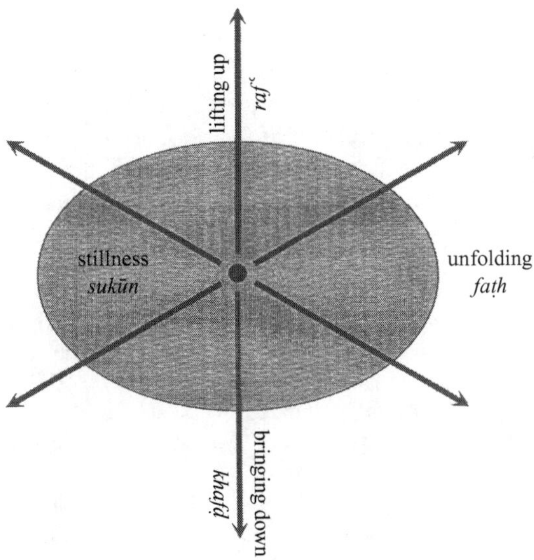

Fig. 2.15 Diagrammatic representation of the formation of the word in Arabic.

The Tree of Being

Kun (Be!) was God's first uttered word, and *kawn* (the world) was the immediate outcome of this utterance. Ibn ʿArabī's treatise *Shajarat al-Kawn* (The Tree of Being) is a fascinating exposition of his mystical reflections on the relationship between *kun* and *kawn*, the command and the outcome, the *word* and the *world*. Among the poetic imageries he constructs is the correspondence between the spatial structure of the human presence (the three-dimensional cross) and the "tree" of realities that grows from the "seed" of the divine word *kun*. In the *Taʿrīfāt*, a dictionary of Sufi terminology, al-Jurjānī defines the term *shajara*, "tree," as "the Universal Man who governs the structure of the Universal Body." The Arabic term *shajara*, "tree," literally means "every plant that stands vertically with a trunk." It derives from *tashājur*, "fighting," "quarrel," and "opposition." Sufis identify the notion of the tree with that of the Universal Man because both embody the pattern of the three-dimensional cross, which expresses notions of both verticality and opposition. The trunk represents the vertical axis, and the branches represent the two horizontal axes.[175] The seed whence the tree grows corresponds to the center, the heart of Universal Man, which is the place where all complements are united and all opposites are reconciled. The Sufi master Abū Saʿīd al-Kharrāz was once asked, "Whereby do you know God?" He replied, "By the fact that he is the *coincidentia oppositorum.*"[176]

Kun is the imperative of *kawn*, which means "cosmos" or "universe," "the world of becoming"; *kawn* also means "coming into existence" and is "used as a noun for 'existence' as a whole, and so the 'universe' as containing all existing things."[177] *Kun* is the principle of *takwīn* (formation), the divine order that can be interpreted as "become" or "come into existence." So the Tree of Being is nothing other than the cosmic tree, and the seed whence it grows is the divine Essence. In *Shajarat al-Kawn* Ibn ʿArabī writes:

> I have looked at the universe *(kawn)* and its design *(takwīn)*, at what was concealed *(maknūn)* and its inscription, and I saw that the whole universe *(kawn)* was a tree, the root of whose light is from the seed 'Be!' *(kun)*. The *K* of the creation *(kawniyya)* was fecundated by the seed of "We created you" (56:57), from which was formed the fruit of "We have created every thing by measure" (54:49) . . .
>
> The first things to grow from this Tree, which is the seed of *kun,* were three shoots. One shoot thereof went to the right; this was "the fellows of the right hand" (56:27). Another shoot went to the left; this was "the fellows of the left hand" (56:41). And yet another shoot, well-balanced in shape, went straight up in a rectilinear way, from which were "the preced-

ers," "those who draw near" (56:11). As it became firm and high reaching, there came from its high and low branches the worlds of meaning and form. What came from the external bark and visible covers was the world of earthly kingdom. And what came from inner cores and concealed meanings was the world of heavenly kingdom. And what came from the sap that runs in its arteries and veins whereby its growth, living, and rising occur, its flowers blossom, and its fruits ripe, was the world of dominating power, which is the secret of the word *kun*.

Then God set a wall around the tree, determined its limits, and drew its forms. Its limits were the directions; they were up and down, right and left, before and behind. So what was highest was its upper limit, and what was lowest was its lower limit. As for its forms they were the spheres, the planets, the angels, the rules, the effects, and the people. So he rendered the seven layers as the leaves sought for their shade, the shining planets as the flowers in the horizons, and the days and nights as two different garments: one was black worn to be veiled from sights, the other was white worn to appear unto those with insights . . .

When the trunk of this tree and its branches stood firm, its two limits met, as its end reached unto its beginning: "Unto your Lord is its termination" (79:44) to its initiation. For whatever begins with *kun* (Be!) it ends with *yakūn* (will be). Thus no matter how many its branches are, and of how many kinds it may be, its origin is one, the seed of the word *kun*, and its end will be one, the word *kun*.[178]

The Geometry of Being

In *Inshāʾ al-Dwāʾir* (The Construction of Circles) Ibn ʿArabī provides a two-dimensional diagram, geometrizing the basic structure of being. The diagram illustrates the relationship between the primordial, divine, and human presences, on which the presence of the word can also be mapped. The primordial presence is represented by the "whiteness," the nondifferentiated background against which the diagram projects. The divine presence is represented by an all-encompassing circle, defining the outer limit of the circle of the human presence, which in turn defines the outer limit of the circle of the world. Mediating between the divine presence and the world, the human presence translates the original unity into the fundamental quadrature of being.

From Ibn ʿArabī's two-dimensional diagram of simultaneous unfolding we can reconstruct the process of universal manifestation in spatial terms. The spatial diagram illustrates the principles of *centrality*, *axiality*, *circularity*, *triplicity*, and *quadrature*, synthesized in one diagram to represent symbolically the underlying order of being. This pattern spatialises the realities of the three

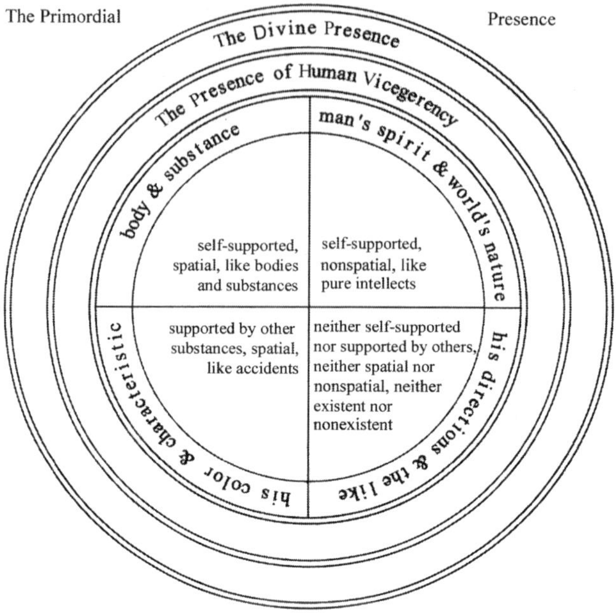

Fig. 2.16 The fundamental order of being according to Ibn ʿArabī *(Inshā' al-Dawā'ir).*

manifest presences. Although the order of this pattern is revealed in each one of these manifest presences, specific aspects may be taken to represent each presence. Centrality and circularity, in that they reflect the order of unity and multiplicity, may be taken to designate the divine presence, whereas triplicity and quadrature, in that they reflect the three dimensions and the six directions, are taken to designate the human presence. As the creative instrument, the presence of the word mediates between the divine and the human presences by expressing the realities of both. All presences coincide in the central point, the expression par excellence of *coincidentia oppositorum.*

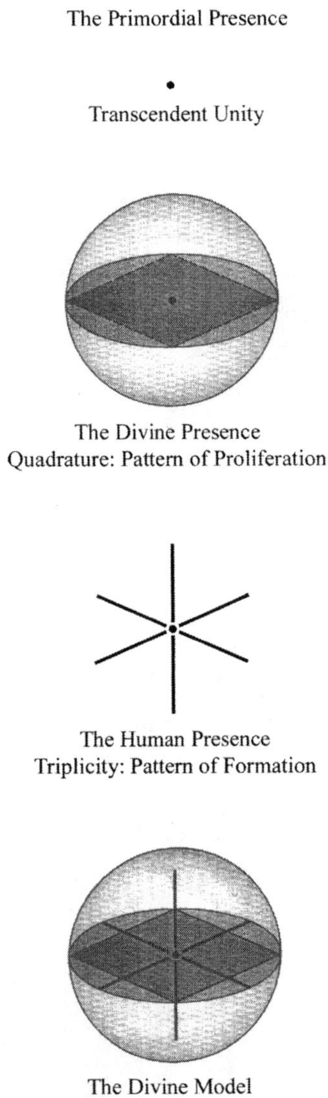

Fig. 2.17 Diagrammatic representation of the geometry of being.

Chapter 3

Cosmic Order

The Original Idea

In *ʿUqlat al-Mustawfiz* Ibn ʿArabī asks us to consider the situation of a person seeking shade and protection, who thought of the idea of a canopy. To build the canopy, however, he first had to prepare the ground and lay down the foundations. In seeking shade and protection, the foundations are the last thing to be thought of yet first to exist. The canopy, by contrast, is the first thing to occur in the mind but last to exist. This is the situation of the world, Ibn ʿArabī says. When God thought of revealing his "hidden treasures," the first thing that occurred in his mind was the idea of humanity. To fulfill this idea, he first had to bring the entire world into existence to form the foundation for human existence. Although last in existence, humanity was the original idea.[1] Humanity could not have existed without the world, just as the canopy cannot stand up without the foundations. And just as the foundation alone without the canopy is meaningless, for it provides neither shade nor protection, so likewise the world without humanity is purposeless, for it lacks the core being for whose purpose it was brought into existence.[2] The celebrated thirteenth-century Sufi Jalāl al-Dīn Rūmī restates Ibn ʿArabī's idea in a poetic manner, drawing our attention to the fact that the outward appearance of things often conceals the inner reality. He writes:

> Externally, the branch is the origin of the fruit; intrinsically the branch came into existence for the sake of the fruit.
> Had there been no hope of the fruit, would the gardener have planted the tree?
> Therefore in reality the tree is borne of the fruit, though it appears to be produced by the tree.[3]

The Sufis along with most premodern Muslim thinkers advocate the view of a purpose-built cosmos designed by God for the accommodation of humankind. Man is at once the center, the model, and the ultimate aim of existence. The ontological correspondence between man and the cosmos was complex and multilayered. It was conceived and presented in a variety of ways in premodern Islamic sources, although the structural core concerning the three-dimensional cross was consistent. Texts such as, for example, the Ikhwān's *Rasāʾil*, Ibn Ṭufail's, *Ḥayy bin Yaqẓān*, Ibn ʿArabī's *al-Tadbīrāt*, and al-Jīlī's *al-Insān al-Kāmil*, reveal rich and sophisticated conceptions underpinned by a firm belief in a universal *order* and structural resonance among the various levels of being. This was not peculiar to the Islamic tradition, of course. In fact the term *cosmos*, from Greek *kósmos*, denotes the idea of "order" and "ornament," meaning the universe as an ordered and ornamented whole. The Arabic equivalent, *kawn*, as already discussed in the Tree of Being, designates the "cosmos" as an embodiment of the metaphysical order. "Cosmic formation" *(takwīn)* refers to the materialization of the immutable essences *(al-aʿyān al-thābita)* in the form of the external essences *(al-aʿyān al-khārijiyya)*, revealing the last three states in al-Hindī's hierarchy: the world of spirits, the world of similitude, and the world of bodies. These worlds correspond to the three modes of cosmic existence: spiritual *(jabarūt)*, angelic *(malakūt)*, and human *(nāsūt)*.

In the metaphysical order, the human presence was presented as mediating between God and the world. This is as far as the designative mode of creation *(taqdīr)* is concerned. In the cosmic order, it is the cosmos that mediates

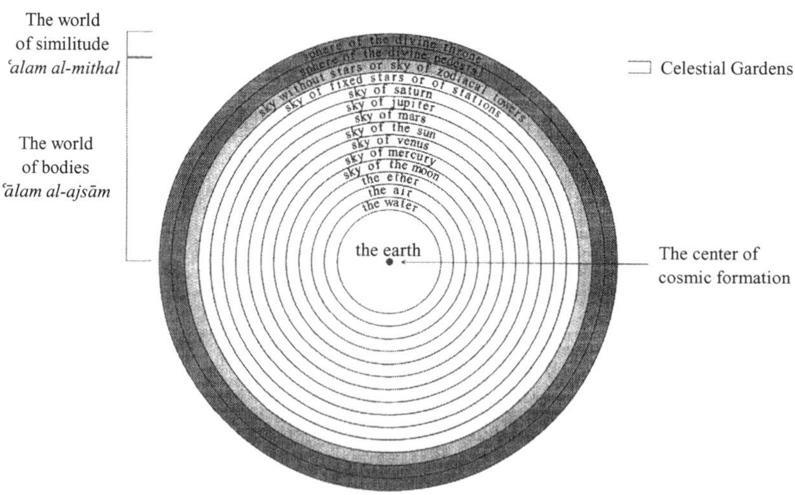

Fig. 3.1 The geocentric cosmos and domains of being according to Ibn ʿArabī.

between God and man, as far as the productive mode of creation *(ījād)* is concerned. The patterns of universal manifestation project into the realm of existence through the production of cosmic forms *(al-ṣuwar al-kawniyya)*. Acting as a link between God and man, the cosmos comprises the formal, imaginable, and communicable vocabularies, which constitute the alphabet of the language of symbolism. By means of this alphabet human imagination is able to function, as already discussed, and by means of the governing order one is able to retrace the geometry of existence according to which the world is fashioned.

In this chapter I will trace the order of divine realities discussed in the preceding chapters at the various levels of cosmic manifestation, focusing on the cosmograms presented in Ibn ʿArabī's *Futūḥāt*. These cosmograms geometrize the cosmic structure at the spiritual, angelic, and human levels of being, revealing the main elements of cosmic landscape that embody the metaphysical order in many and varied forms.[4] Thus the analysis of cosmic order deals with these various modes of embodiment, illustrating the way in which the formless, yet intelligible, relationships between the divine realities are translated into imaginable, spatio-temporal expressions. It demonstrates how all levels of cosmic hierarchy are gathered by the *niẓām*, at once "order" and "thread," of the divine realities that ties all manifestations together and everything back to their original source.

Creative Breathing

The utterance of the creative order, Ibn ʿArabī says, coincides with both the exhalation of the divine Breath *(al-nafas al-ilāhi)* or the Breath of the Compassionate *(al-nafas al-raḥmānī)* and the manifestation of the world.[5] Through the agency of the Breath the manifestation of the world becomes synonymous with the self-disclosure of the Absolute.[6] Self-disclosure, like creation, has two distinct phases: first, the essential Self-disclosure *(al-tajallī al-dhātī)*, wherein the Absolute manifests as immutable essences; and second, the sensuous Self-disclosure *(al-tajallī al-asmāʾī)*, wherein the Absolute manifests as external essences.[7] As an inward act that occurs within the divine Self or Consciousness, the essential Self-disclosure does not project outwardly in an *otherness* differentiated from the *sameness* of the Essence. The immutable essences manifested by this determining act are nothing other than the names and attributes of the essence before externalization. Otherness occurs in the sensuous Self-disclosure when these names take on forms, through God's exhalation of "the first dense, transparent, luminous mass," the "compassionate vapor" *(al-bukhār al-raḥmānī)*, that is, the divine Breath.[8]

What is this compassionate vapor? And why did God exhale it? The realities of the world, Ibn ʿArabī explains, were within the Essence in a state of anxiety and distress *(karab)*, crying out for externalization.[9] Bālī Efendī, the sixteenth-century Sufi, compares this to the holding of one's breath within and the associated "painful sensation of extreme compression" as the breath seeks an outlet. Only when one breathes out does this compression cease. Similarly, Efendī says, "the Absolute would feel the pain of compression if it did not bring into existence the world in response to the demand of the Names."[10] This state of distress caused "the sadness of the primordial solitude" that made God yearn to reveal himself: "I was a hidden Treasure, I yearned to be known. That is why I produced creatures, in order to be known in them."[11] To alleviate this distress *(karab)* God breathed *(tanaffas)*, and by breathing he externalized the inner realities, compassionately responding to their cry. The Breath of the Compassionate, as al-Qāshānī puts it, brought out potential beings *(al-mumkināt)*, which had they remained in nonexistence they would have caused the "distress of the Compassionate" *(karab al-raḥmān)*.[12] Thus the attribute of compassion characterizes the act of bringing into existence the forms of the world, and that is why the creative medium is called the "Breath of the Compassionate" *(al-nafas al-raḥmānī)*.[13]

Breathing involves a repetitive act of inhalation and exhalation. Sufis use this metaphor in their concept of perpetual 'renewal of creation' *(tajdīd al-khalq)*. They say that by the inhalation and exhalation of the divine Breath all cosmic forms contained in the Breath are constantly manifested and reabsorbed, ceaselessly renewing the creation at every moment. The concept of the 'divine Breath' also forms the foundation for alphabetical symbolism, already discussed. Al-Qāshānī says that as God attributes to himself the Breath, it is necessary to attribute to him also all of what the Breath involves, like breathing forth *(tanfīs)* and articulating the forms of the letters and words that, in this case, are the cosmic words *(al-kalimāt al-kawniyya)*.[14] Through the Breath meanings and letters, as spirits and forms, become fused together. The forms of the world receive the animating spirit from the Breath of the Compassionate in the same way letters receive meanings the moment they are pronounced.[15]

Ibn ʿArabī asks those seeking to understand the nature of the divine Breath to consider the world. All is contained in the divine Breath like the day in the morning's dawn, he says, meaning that the world actualizes the forms potentially disseminated in the Breath, in the same way the day brings about all the events already ordained in its first moment, the dawn.[16] In philosophical terms, the divine Breath is the original medium through which potential beings were externalized, bursting out from the inwardness of formless potentiality into the outwardness of formal actuality. It is the "substance of the world" *(jawhar al-ʿālam)*, wherein are latent all the possibilities of formal manifestation.[17] The Breath equates the prime matter *(al-hayūlā al-ʾūlā)*, which englobes all the forms of the world, representing, in the Ikhwān's terms, the transcendent substance of all divine artefacts.[18] The Breath is to the world what the intelligible point is to geometry and what the "ink" is to al-Ghazālī's archetypal exemplar: the source wherein all possibilities are fused together as a nondifferentiated totality. It is to God what the whiteness of a blank sheet is to the architect: the unformed *materia* that is susceptible of receiving all kinds of forms. The divine Breath is at once the creative medium and the necessary substantial support for all creations.

Breathing as Imagining

In response to the question "Where was our Lord before creating his creatures?" the Prophet is reported to have said: "in a *ʿamā'* with no air either above or below."[19] The Arabic term *ʿamā'* literally means "thin and subtle cloud." According to Ibn ʿArabī, it refers here to the divine Breath. The primordial "Cloud" is thus the first form the Breath took on externally and within which God then differentiated the forms of the world.[20] In the context of the geometrical and alphabetical symbolism, the Cloud can be seen as the cosmic equivalent of the circle and the *alif*, that is, the first affirmatively conceived reality and the first qualified form of unity. It is the governing form within which the realities of the world are delivered from potency into actuality, from formlessness to formal existence. Ibn ʿArabī considers the Cloud to be the first existential condition *(ẓarf)* that supported God's external being *(kaynūnat al-ḥaqq)*, while at the same time identifying it with absolute imagination *(al-khayāl al-muṭlaq)*.[21] The Cloud is identified with the divine imagination because it is viewed not only as a passive substance capable of receiving all forms but also as active agent that gives beings their forms.[22] It is thus the means whereby God projected forth the essences of potential beings as cosmic, imaginable forms, and the instrument whose function is to actualize the transcendental patterns of divine realities in the harmonized form of the cosmos.

By identifying the Cloud with absolute imagination Ibn ʿArabī presents divine breathing as an act of imagining. Unlike human imagining, he argues, divine imagining occurs from without and not from within the Essence. This is to say that God produced the world the moment he imagined it and not according to an eternally imagined model *(mithāl)*. And prior to their existence in the Cloud, the forms of the world did not exist as such in the divine Self, nor has God imagined them in his Mind prior to their production. As immutable essences, they were *known* as they are and as they would be when formally produced but not imagined.[23] The divine imagining of the forms of the world coincides with producing them through the Breath, hence the conflating of the divine acts of breathing and imagining.[24] Peculiar though it may sound, this conception is fundamental to Ibn ʿArabī's approach to resolve the perennial philosophical problem of the eternity *(qidam)* and newness *(ḥudūth)* of the world. Through breathing-as-imagining Ibn ʿArabī attempts to reconcile the eternity of the world as immutable essences with the Islamic dogma of *creatio ex nihilo*.[25]

To resolve this philosophical dilemma, Ibn ʿArabī begins by making a clear distinction between form *(ṣūra)* and meaning *(maʿnā)*, imagining and knowing, as already discussed in chapter 1. Forms embody formless meanings, and as such they are accessible by human imagination. "The forms, insofar as they are forms," he says, referring to the cosmic forms, "are the imaginable, and the Cloud, in which they are manifested, is the imagination."[26] Thus viewed, Ibn ʿArabī's forms are not permanent, Platonic models in whose likeness things are made but are rather the things themselves. There are pure, spiritual forms just as there are sensible, gross forms and intelligible, subtle forms. Together they constitute the cosmic forms that embody the formless immutable essences. In Ibn ʿArabī's scheme of the creation, "cosmic" and "formal" are therefore synonymous terms. Meanings, on the other hand, are accessible by the intellect and can be known without necessarily being imagined. The original meanings are none other than the immutable essences.[27] Accordingly, the imaginable forms that Ibn ʿArabī speaks of as existing in the Cloud or the detached imagination are different to the knowable immutable essences, which "have not smelt the fragrance of existence," residing as they are in the divine Self.

The distinction between meaning and form, knowing and imagining, is consistent with Ibn ʿArabī's conviction that knowledge is not the knower imagining the form of the known, as already discussed. He finds support for this in the divine name *badīʿ*, "originator" or "innovator," mentioned in a verse that speaks of "the originator *(badīʿ)* of the heavens and the earth" (2:117).[28] This name derives from *ibdāʿ*, which means "to bring forth something original, novel, unprecedented," and of which the term *bidʿa* means "originality," "novelty," and "heresy." Commenting on the above verse, Ibn ʿArabī says that the creation of the heavens and the earth is associated with the name *badīʿ* because they are created according to no preceding "model," "likeness," or "form" *(mithāl)*. Had the form of the cosmos been identical with the immutable essences in the nonexistence, God would not have been *badīʿ*, for he would have been creating according to the form already present in his knowledge, and there would be no *creatio ex nihilo*.

> God says: "The originator of the heavens and the earth" because they were created according to no preceding model. The first thing God created was the Intellect, that is, the Pen *(al-qalam)*: it is the first original creature *(mafʿūl ibdāʿī)* manifested from God-most transcendent. And every creature created without a preceding model *(mithāl)* is original *(mubdaʿ)*, and its creator is its originator *(mubdiʿ)*. So if knowledge is conceiving the form of the known, as some people maintain in the definition *(ḥadd)* of knowledge, that creature would not be original *(mubdaʿ)*, because it has in the soul of the one who originated it a model,

according to which he brought it into existence. To maintain this definition of knowledge would mean that that which is in God's Self has never ceased to be necessary being *(wājib al-wujūd)* and that God did not originate *(ibtadaʿa)* it in himself, as does the innovator *(al-muḥdith)* when he originates, nor has anything been brought into existence but according to the form, which exists in the Self of the form giver *(al-muṣawwir)* for [the sake of things to be in] its likeness not for its own sake, for [God's Self] is not the place of what he creates. It follows that God is not *badīʿ* (according to those who maintain that knowledge is the form of the known imprinted in the soul of the knower); but he is. So he has in his Self no form of what he originates, nor has he conceived of its form [before originating it]. This is a problematic matter. Among the knowable matters *(maʿlūmāt)* there are things that can be formalized and others that cannot, though they are knowable; hence, the definition of knowledge is not conceiving the form of the known. And so likewise is the one who knows; he could be amongst those who are able to conceive of forms, being endowed with the imagining faculty, and could be amongst those who know without being able to formalize, being incapable of giving form. Thus, [for God] form giving is an act that occurs from without *(min khārij)*, and he does not receive within his Self what he forms *(ṣawwara)* from without, but he knows it. And know first that origination *(ibdāʿ)* is not possible except with forms *(ṣuwar)* in particular, because they can be created and can, therefore, be originated. As for meanings *(maʿānī)*, none of them is originated *(mubtadaʿ)*, because they cannot be created nor can they be originated, though they can be intellectualized as being essentially immutable.[29]

The "Cloud" and Cosmic Forms

Ibn ʿArabī's elaboration on the nature of the forms contained in the Cloud adds further clarity both to the distinction he makes between form and meaning and to the relation he establishes between the primordial Cloud and the world of detached imagination. Commenting on the verse "Everything will perish save his Face *(wajhihi)*" (28:88), Ibn ʿArabī explains that *his* in "his face" (the pronominal suffix *hi* in *wajhi-hi)* can be understood as referring to the "thing" in "everything." The verse would then read as "Everything will perish save its face." Similarly, in the prophetic tradition "God created man in his Image *(ṣūratihi)*," the same pronoun may also refer to "man," meaning God created man in man's own image. Understood in the alternative sense, Ibn ʿArabī considers the *form* of a thing to be its perishable aspect revealed in the Cloud, whereas its "face" to be its imperishable reality. He explains:

Then he caused to exist in the Cloud all the forms of the world, about which he said, "It will perish," that is, in respect of its forms, "save its face," that is, in respect of its reality it will not perish. For the *hā'* in *wajhihi* refers to the

> "thing." So in relation to the forms of the world, "everything will perish," but in relation to its realities, the world will not perish, nor is it possible to perish. If the form of man perishes, for example, and there remains no trace of it in existence, its reality, which is identified by, and is identical with, man's definition *(ḥadd)*, would not perish. We say that man is a "rational animal" *(ḥayawān nāṭiq)*, and we do not refer to his being existent or nonexistent, because this reality has never ceased to be his even if there were for him no form in existence.[30]

Within the primordial Cloud God unfolded the forms of the entire world, highest and lowest, subtle and dense, spatial and nonspatial. Ibn ʿArabī illustrates these forms in a series of diagrams, which show in a hierarchical order both the supra-natural and the natural worlds with all the cosmic levels they comprise. In the following I shall examine some of these diagrams in the same sequential manner Ibn ʿArabī follows, though he indicates that they should be seen as one diagram, in which the simultaneous existence of the elements would enable a better appreciation of their proper relationships.[31]

The World of Command

"To him belong the creation and the command" (7:54), the Quran says, introducing an important duality that underpinned premodern Islamic cosmology. Muslim theologians interpreted the verse as referring to two distinct worlds: the world of command *(ʿālam al-amr)* and the world of creation *(ʿālam al-khalq)*. This duality is consistent with the then prevailing Neoplatonic conception of the sensible and intelligible or physical and metaphysical division of reality. Ibn ʿArabī's first diagram illustrates the metaphysical world of command.[32] Signifying authority and control, *ʿālam al-amr* designates the realm where the immutable laws governing all forms of worldly existence are set. In al-Hindī's hierarchy, it is the fourth state of Being, the world of spirits *(ʿālam al-arwāḥ)* that comprises the simple, cosmic models, in the likeness of which things are fashioned. It is the highest level in the hierarchy of cosmic manifestation, the level where God revealed his design of the world through the luminous traces the Pen inscribed upon the Tablet.

Ibn ʿArabī's rather curious diagram of the world of command shows the first stage of formal articulation within the primordial Cloud. It shows the Cloud in the form of an encompassing circle, the circumference of which is marked by the thirty stations of the angels ecstatic with love *(maqāmāt al-malāʾika al-muhayyama)*. These encompass the quadrature of the Pen, the Preserved Tablet, Nature, and Matter, with each assuming distinct geometrical shape. The Pen is identified as the First Intellect *(al-ʿaql al-awwal)*, and the

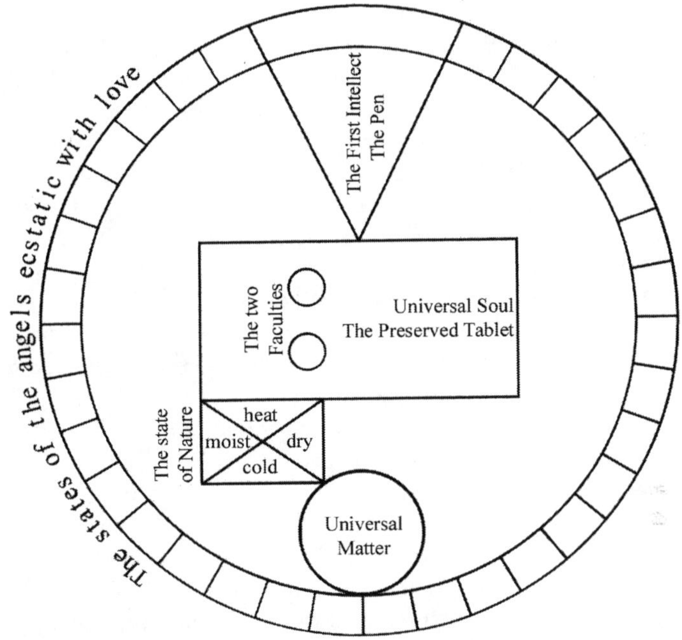

Fig. 3.2 The form of the "Cloud" revealing the world of command according to Ibn ʿArabī *(Futūḥāt)*.

Preserved Tablet as the Universal Soul *(al-nafs al-kulliyya)*. Within the Preserved Tablet two smaller circles are shown, representing the Soul's intellectual and practical faculties.[33] Next to the Preserved Tablet, the state of Nature *(martabat al-ṭabīʿa)* is shown as a rectangle divided diametrically into four parts. In these divisions the four principles of Nature are arranged in two antinomical pairs: heat/cold and dryness/moistness. Next to the state of Nature, the Universal Matter *(al-hayūlā al-kull)* appears in the form of a circle analogous to the encompassing circle of the Cloud.[34]

Ibn ʿArabī's diagram *re-presents* the already-discussed quadrature of the Intellect, Soul, Nature, and Matter in a new way. Comparing this with the diagram discussed in the chapter on the divine presence helps understand the difference between the two modalities. At the divine level these elements were formless. The divine geometry emphasizes a particular configuration of relationships and an inherent propagative order. It reveals the symmetrical relationships the first created quadrature bear to the original divine attributes and the governing patterns they inscribe. The cosmic geometry, by contrast, emphasizes their distinct forms, functions, and existential context. The Intellect, a formless reality at the divine level, assumes the form of an angel ecstatic with love at the cosmic level and becomes distinguished, unlike the rest of the angels in its class, by the unique capacity of intellectualizing both itself and its creator. The Soul also takes on the form of an angel ecstatic with love and becomes equipped with the theoretical and practical capacities. Nature, as a state, appears through its four generative forces. While Matter, the substance of the physical world, assumes a circular form specifically conditioned for spatial determinations.

Ibn ʿArabī's visualization of the world of command has its roots in the prevailing mythology and reported prophetic traditions. One tradition says: "God has a white earth in which the sun takes thirty days to cross the sky, and each of these days is thirty times longer than the days of the lower world. That earth is filled with creatures who do not know that God has been disobeyed in the earth or that He has created Adam and Iblis."[35] This is the prophetic reference to the angel ecstatic with love, who, according to Ibn ʿArabī, are the first luminous bodies God created. Apart from the Soul, which was created through the agency of the Intellect, these luminous bodies were created without the agency of other beings. They are nonspatial bodies lying above the ruling of Nature. Being the most rarefied form of bodies, they define the universal boundary, the transitional zone, between the formal and the formless. They form the circumference of the Cloud, the outer limit of the universe. Created from the light of the divine Majesty these principial spirits are enraptured with God's beauty for they are exposed to nothing else.[36] Among them only the Intellect and the Soul are charged with responsibilities toward the created world.[37]

The metaphysical order underpins the various embodiments that take place at this cosmic level as revealed in the triangle of the Intellect, the rectangle of the Soul, the divided square of Nature, and the circle of Matter. Ibn ʿArabī does not explain the logic of this diagram's asymmetrical composition. Despite the overall circular form, there seems to be a sense of vertical hierarchy in the composition, suggesting a vertical reading of the diagram. The pointedness of the Pen above the Tablet establishes their ranking, while their connection with the physical world, that is to unfold with the Universal Matter, is mediated by the state of Nature. Circularity, as an expression of unity, totality, and firstness, appears in the form of both the Cloud and Universal Matter. The form of the Cloud, as the cosmic expression of the first qualified form of unity, is mirrored in the form of Universal Matter. But whereas the circle of the Cloud encompasses both the metaphysical and the physical, the Universal Matter is specially conditioned for physical manifestations.

Ibn ʿArabī differentiates two levels of Nature: grand and limited. "Nature," he says, "is the most deserving relation to be identified with the

Real because everything else was manifested by it."[38] It is the Breath that pervades the world, ruling over all forms including the Intellect. Here Ibn ʿArabī is referring to the grand Nature *(al-ṭabīʿa al-ʿuẓmā)* that is inherent within God's creative medium, the Breath.[39] The two levels of Nature are similar, however, in the way that a mother and a daughter are capable of maternity and progenitive production. The state of Nature shown in the above diagram is the "daughter" of the grand Nature.[40] As Nature has no essence, however, it is traced through the substance within which it reveals its effects. The grand Nature is, therefore, identified with the Cloud, and the "daughter" with the Universal Matter.[41] The Cloud, the first luminous mass, is the primal foundation of all, whereas Universal Matter, also *al-habāʾ*, is the substance of the spatio-temporal world. *Al-habāʾ*, the Jurjānī's *Taʿrīfāt* says, is "the very substance in which God unfolded the bodies of the world."[42] Thus understood, *al-habāʾ* is not a prime substance in the sense of pure potency, but rather a relative or secondary substance that is determined in accordance with the special demands of the spatio-temporal conditions of existence.[43] In this sense, Universal Matter corresponds to *materia secunda* in the scholastic philosophy, whereas *materia prima* equates Ibn ʿArabī's *ʿamāʾ* and the Ikhwān's Original Matter.[44]

Structured upon quadrature, the state of Nature mediates between the exemplar set in the Tablet and its embodiment in the Universal Matter. Yet Nature, as Ibn ʿArabī explains, has no inherent knowledge of its own, nor has the instrument of acquiring knowledge. As an active force, it acts under the directives of the Soul, and quadrature is the underlying structure of its modus operandi. Although it functions by means of the four generative principles, only two of these are active. This is because it is subject to the Soul's determination, Ibn ʿArabī explains, and the Soul has only two faculties: the theoretical and the practical. So Nature's two active instruments come from the Soul's inherent dual structure. They work together: the Soul provides knowledge, and Nature acts. This reflects the understanding that the natural laws governing worldly existence alone cannot explain the reality of things. The laws of nature constitute the object of the science of causes *(ʿilm al-asbāb)*, whereas the inner meanings can only be accessed through the science of realities *(ʿilm al-haqāʾiq)*.

Nature's two active principles are heat and cold. Heat causes dryness, and cold causes moistness. Dryness and moistness are thus passive outcomes in relation to heat and cold. They are also in opposition just as are heat and cold.[45] Heat negates cold, and dryness negates moistness; hence they cannot naturally mix. Their contrasting qualities, however, enable a particular pattern of productive synthesis, wherein quadrature remains the underlying order. Active and passive forces of Nature join in four possible combinations only, producing the four natural elements: fire, air, earth, and water.[46]

Heat			**Cold**		
Dryness	+	Moistness	Dryness	+	Moistness
Fire		**Air**	**Earth**		**Water**

The active-passive interaction applies to the Pen and the Tablet as well, as both are subject to the ruling of the grand Nature. The Pen represents the essential, active pole of manifestation, and the Tablet represents the substantial, passive pole. The Pen, as the "cosmic refraction" of the primordial word,[47] embodies the triplicity of the creative command "Be!"; whereas the Tablet, as the cosmic book, actualizes the command, materializing the quadrature of the *arkān*—fire, air, water, and earth. The Pen corresponds to the productive triplicity of formation, whereas the Tablet corresponds to the designative quadrature of proliferation. A reported prophetic tradition says that the first thing God created was the Pen, whose length equaled the distance between heaven and the earth. He then created the Tablet, whose length extended between heaven and earth, and its width stretched from east to west.[48] The Pen, thus viewed, signifies axiality, corresponding to human spirituality and unique upwardness, to the verticality of the *alif*, to the trunk of the Tree of Being, and to the vertical axis of the three-dimensional cross. The Tablet signifies the principle of horizontality, corresponding to the human corporeality, to the letter *bāʾ*, to the branches of the Tree of Being, and to the two horizontal axes of the cross. The Tablet also corresponds to the circle, reflecting the divine presence, and the Pen corresponds to the point, reflecting the Essence. Just as the Essence, under the "pressure" of the realities, exhaled the Breath, manifesting the forms of the world, the mother point, wanting to reveal its hidden treasures, gave birth to the multitude of potential beings, and the seed of "Be!" after fecundation generated the cosmic tree, so likewise the Pen, after looking toward God with "a look of reverential fear *(hayba)*,"[49] burst open, the ink *(midād)* of existence flowed, and the exemplar of the world was transcribed.

The World of Creation

Zooming in on the circle of Universal Matter, we cross the threshold from the world of command *(ʿālam al-amr)* into the world of creation *(ʿālam al-khalq)*, from the metaphysical to the physical. As an imaginary substance specially conditioned for physical manifestations, Universal Matter is an intelligible reality, or an agency, with no essence *(maʿqūl ghayr mawjūd al-wujūd al-ʿaynī)*.[50] It is recognizable through the forms that unfold within it. These are the world of creation represented by the main cosmic features: the divine

Throne, the Footstool, the celestial Gardens, the heavens and the earths. These cosmic forms are subject to the governing effects of Nature and are therefore distinguished from the supra-natural world that lies above them.[51]

The Throne and the Footstool

The natural world in Ibn ʿArabī's scheme is the world of synthesis, and the Throne is the first composite form that marks the threshold into the domain of complex cosmic entities.[52] But composition is not in the material sense yet, for the Throne and the Footstool are not literally spatial entities. Synthesis refers here to the mediating realities involved in their production. The Throne is considered to be a composite form because its production involves four realities: Nature, Universal Matter, Universal Body, and Circularity. God first brought Universal Matter into being, which was then transformed into the Universal Body upon receiving the spatial qualities of length, breadth, and depth. Nature then conditioned this Body by governing its possibilities. Circularity was the first form this Body received, so there was the sphere *(falak)*. God called this sphere the "Throne" and as the all-Compassionate he rested upon it.

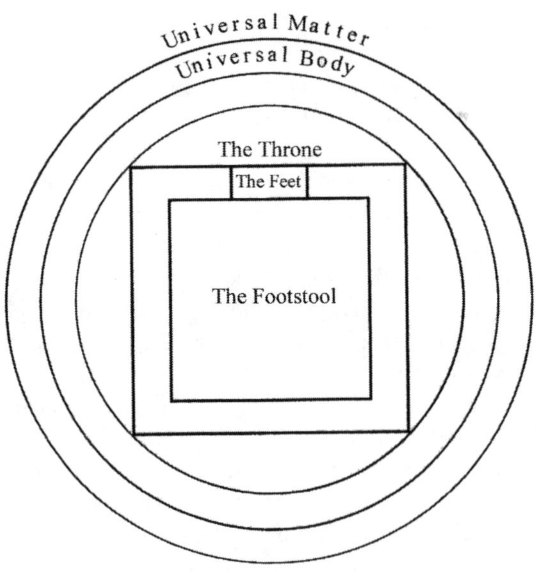

Fig. 3.3 The divine Throne and the Footstool according to Ibn ʿArabī *(Futūḥāt)*.

The Quran describes the Throne as the divine seat, "The all-Compassionate sat himself upon the Throne" (20:5), and his Footstool *(kursī)* as "encompassing the heavens and the earth" (2:255). It also refers to the bearers carrying the Throne and the angels surrounding it. *ʿArsh*, "throne," has two related meanings: the "kingdom" over which a king reigns and the "seat" *(sarīr)* upon which he sits. The Throne was accordingly seen as the whole physical world as well as the cosmic seat upon which God rests. Citing Ibn Masarra, Ibn ʿArabī says that the lifted Throne *(al-ʿarsh al-maḥmūl)* is none other than the divine kingdom *(al-mulk)*.[53] The bearers can thus be interpreted as the basic structure and governing laws of the kingdom as well as the pillars that support the Throne.[54] As for the angels, Ibn ʿArabī explains that God created them from the lights of the Throne, which is described in a tradition as being created from divine Light. From these angels God selected four bearers to carry the pillars of the Throne.[55]

As the threshold into the physical world, the Throne and the Footstool define a transitory domain that is neither spatial nor nonspatial. Ibn ʿArabī attributes to the Throne sensible, spatial characteristics, describing it as "a seat with four corners and four faces." Although the four corners are the "original pillars," they are not the sole supports of the Throne. "In each of the Throne's four faces," Ibn ʿArabī adds, "there are many pillars equally distributed." The Throne is also made hollow to contain the physical world.[56] With reference to the verse, "The angels and the spirit ascend unto him in a day whereof the span is fifty thousand years" (70:4), Ibn Kathīr says, "The distance between the Throne and the seventh earth is fifty thousand years of travel, and its breadth is fifty thousand years."[57] But despite these spatial characteristics, Ibn ʿArabī warns, the Throne is not *spatial* in the literal sense. To take the spatial characteristic literally, he says, one would face the difficult task of explaining in spatial terms the modes of divine "sitting" upon it and the angels "encircling" it. "If you say that there is, for the angels who are encircling the Throne, no space to move within, since the Throne has occupied the entire vacuum," Ibn ʿArabī explains, "we say, there is no difference between them encircling the Throne and God resting upon it. For that which does not admit spatiality does not admit conjunction and separation."[58] The same applies to the Footstool, which shares with the Throne its subtle modality, for just "as the Compassionate rests upon the Throne, the Feet rest upon the Footstool."[59] By virtue of its subtle, intermediary nature, the domain of the Throne and the Footstool combines the spiritual and physical characteristics of its neighboring domains.

The forms of the Throne and the Footstool crystallize the pattern of quadrature. The four bearers of the Throne correspond to the four creative attributes—Life, Knowledge, Will, and Power—that support the divine presence. Many traditions say that God created the Throne from green emerald and its four pillars from red ruby.[60] The bearers of the Throne are also depicted as four angels whose feet are in the nethermost earth and whose heads are in the Throne. These four bearers have the forms of a man, a bull, an eagle, and a lion.[61] Four celestial rivers are also described as being laid out about the Throne: a river of sparkling light, a river of blazing fire, a river of shining white snow, and a river of water.[62]

The bearers of the Throne, Ibn ʿArabī explains, are the governors responsible for the management of the natural world. They are four pairs of archangels and companion prophets. The angels are Seraphiel (Isrāphīl), Gabriel (Jibrāʾīl), Michael (Mīkhāʾīl), and Ruḍwān, and the prophets are Adam, Muḥammad, Abraham, and Mālik.[63] Each of the four pairs is responsible for one core matter: form *(ṣūra)*, spirit *(rūḥ)*, nourishment *(ghidhāʾ)*, and status *(martaba)*. Seraphiel and Adam are responsible for forms, Gabriel and Muḥammad for spirits, Michael and Abraham for nourishment *(arzāq)*, and Mālik and Ruḍwān for status. Each pair of the Throne's supporters constitutes as it were two complementary aspects: hidden and manifest, spiritual and sensible. This is reflected in the elements they support. Ibn ʿArabī explains that forms are of two kinds: luminous like those of the "angels ecstatic with love," and sensible like those in the natural world (including imaginary forms); spirits are divided into those associated with luminous forms and those with sensible forms; nourishment is also divided into sensible, such as food, and spiritual, such as sciences and knowledge; and the status of every being is polarized into happiness and suffering, which have many sensible and spiritual forms in this world and in the hereafter.

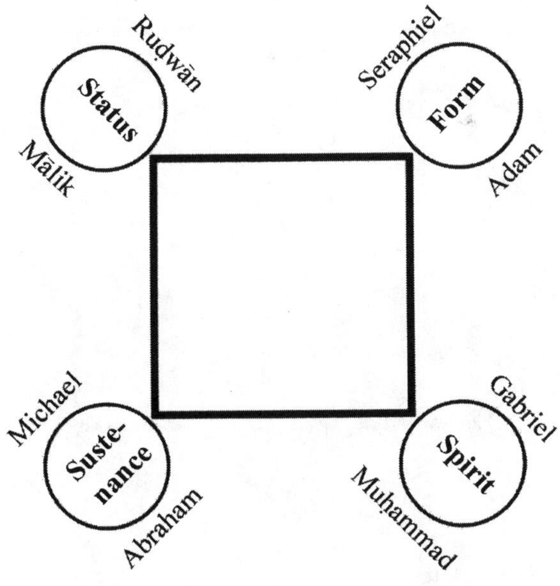

Fig. 3.4 The bearers of the divine Throne according to Ibn ʿArabī.

Although quadrangular in form, Ibn ʿArabī says, the Throne has eight pillars. The other four pillars, for which there are no bearers in this world, mark the middle of each of the Throne's four faces. On the day of the resurrection God will appoint those who shall bear the throne from these pillars, as alluded to in the following verse: "And eight will uphold the Throne of their Lord that day" (69:17).[64] Accordingly, the quadrangular structure of the Throne is tied to the structure of the "first formation" *(al-nashʾa al-ʾūlā)*, supporting existence in the present world. In the "other formation" *(al-nashʾa al-ʾukhrā)*, when the world will be recreated afresh, the Throne will become octagonal as the new four supporters will assume their responsibilities.[65] In this formal transformation the very structure of the Throne will not change, only four new bearers will join the original four. Ibn ʿArabī illustrates geometrically the structure of the divine Throne in the other formation by way of two superimposed squares, a form that reveals at once the original quadrature of the first formation as well as the octagonal structure of the other formation.

"Inside the Throne," Ibn ʿArabī says, "God created the Footstool *(al-kursī)*, square in form, and let his Feet to hang down onto it."[66] *Al-kursī*, the cosmic

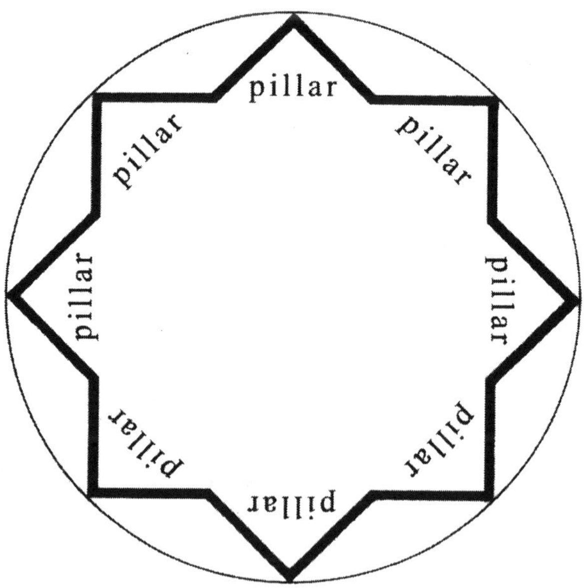

Fig. 3.5 The form of the Throne in the hereafter according to Ibn ʿArabī *(Futūḥāt)*.

container of the heavens and the earth, is seen as the support of the divine Feet. It is quadrangular in form, sharing with the Throne similar features. A tradition says: "Those who bear the Footstool are four angels, each of whom has four faces; their feet are in the Rock below the seventh, nethermost earth."[67] Ibn ʿArabī confirms that the Footstool resembles the Throne, but only with regard to its quadrature, not its pillars. The main difference lies in that the Footstool identifies the level at which the first bifurcation of unity occurs symbolized by the Feet *(al-qadamayn)*. As the Feet hang down onto the Footstool, Ibn ʿArabī explains, "the divine word, which was one on the Throne, is divided."[68] The divine compassion, one at the Throne of the all-Compassionate, splits at the Footstool into compassion *(raḥma)*, symbolized by one foot, and wrath mixed with compassion *(ghaḍab mashūb bi-raḥma)*, symbolized by the other. At the level of the Footstool, the absolute compassion of the Throne becomes relative by associating it with wrath, its opposite. Thus the divine Feet signify the first polarization of unity, the model for all binary oppositions, which fall under either commandment *(amr)* and prohibition *(nahī)* or affirmation *(ithbāt)* and negation *(nafī)*. These binaries govern the physical world. They are the cosmic referents of God's antinomical names and attributes, which are in turn the referents for all opposites in the world, whose spatial expression par excellence are the six arms of the three-dimensional cross projecting from a common center into opposite directions.

The Throne is associated with two ideas: the idea of light *(nūr)* and the idea of spirit *(rūḥ)*. The spirit, "most often figured as a 'center', a 'ray', a 'descent', a 'presence' or 'immanence'," is thought of as residing at the center of the Throne. As the sphere of spheres, the outermost, all-encompassing limit of our world, the Throne forms the "circumference" *(muḥīṭ)* of the body of the world. Signifying totality and integration, the Throne identifies a universal domain of which spirit is the center and light is the matter.[69] Thus viewed, the Throne embodies the principle of circularity, reflecting the pattern of the divine presence, wherein the Essence corresponds to the central spirit, and the names correspond to the encompassing circumference.[70] The process through which the undifferentiated divine Light becomes differentiated by taking on the form of the Throne simulates the process through which the undetermined Essence becomes determined by descending into the state of the first determination. And just as the process of the essential determination continues to distinctively reveal the divine names and to designate the creative quadrature of Life, Knowledge, Will, and Power, so likewise the process of differentiation of the divine Light continues to manifest the angels, who encircle the Throne, from the lights of the Throne *(anwār al-ʿarsh)*, and to designate the supportive quadrature, who are entrusted with the task of carrying the Throne.

Combined with the vertical ray of the spirit, which stands at its center, the quadrature of the Throne constitutes the three-dimensional cross, the pattern of the human presence. The spirit is represented by a vertical ray since it is "the affirmation of Unity in all the degrees of universal Existence," the vertical link that ties all beings to their originator.[71] Numerically, the octagonal order of the Throne corresponds to the number 8, the order of the divine presence: the Essence and the seven principal names—Living, Knowing, Willing, Powerful, Speaking, Hearing, and Seeing. Eight is seen as the first cubical number, which, as previously discussed, corresponds to the three dimensions of length, breadth, and depth. Hence the octad of the Throne also corresponds to the triplicity of the human presence. The triplicity of the three dimensions is the pattern by means of which Universal Matter becomes the Absolute Body, which then receives the forms of the Throne and the Footstool.

The Celestial Gardens

Within the realm of the Throne and Footstool, Ibn ʿArabī locates the celestial Gardens *(al-jinān*, singular *janna)*, the faithfuls' promised abode of eternal happiness. Premodern Islamic sources reveal an interesting debate concerning whether the Gardens are already created or are part of God's scheme of the second creation. The point of the debate is why God would create something useless that he will have to destroy and recreate again at the time of resurrection. Ibn ʿArabī articulates a sophisticated view with regard to this debate, arguing for the existence of hierarchically structured Gardens located in a cosmic domain that will not be subject to destruction and recreation. This domain is bounded by two spheres that God created within the Footstool: the sphere without stars *(aṭlas)*, its upper limit, and the sphere with fixed stars *(falak al-kawākib al-thābita)*, its lower limit. The convex surface of the latter sphere forms at once the ground of the Gardens and the upper limit of the planetary skies, the domain that will be consumed by the fire of Hell in the hereafter.

The *aṭlas* sphere is a "circular, transparent body" that God divided into twelve sections, *burūj*, as alluded to in the following verse: "By the heaven, holding mansions of the stars *(burūj)*" (85:1).[72] It is the sphere of the constellations. The Arabic term *aṭlas* means "effaced" or "obliterated," denoting the idea of perfect homogeneity without any distinguishable features. The *aṭlas* sphere thus forms the homogeneous background onto which the configurations and movements of the planets and the stars are projected. In "The Anatomy of Spheres," the sixteenth- and seventeenth-century astronomer Bahāʾ al-Dīn al-ʿĀmilī considers the aṭlas *sphere* and the *sphere of the fixed stars* to be the scientific terms for the Throne and the Footstool respectively.[73] This might have been so, but as we will see, this does not work with Ibn ʿArabī's cosmological

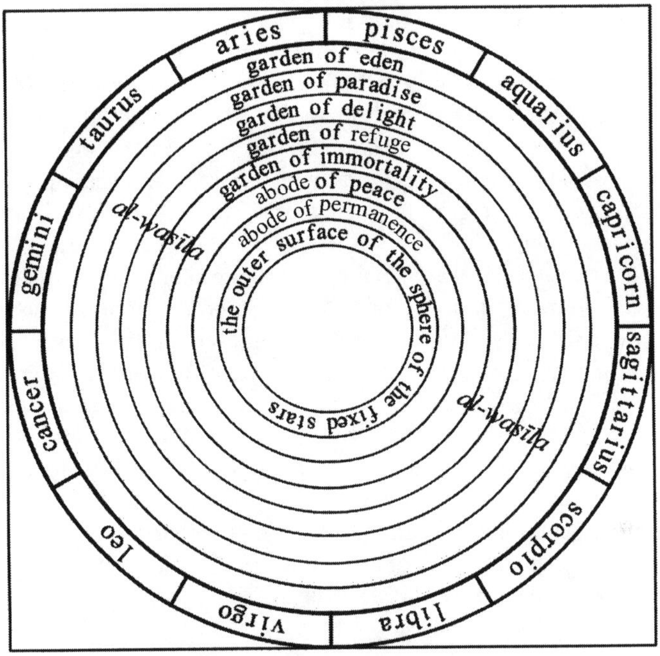

Fig. 3.6 The celestial Gardens according to Ibn ʿArabī *(Futūḥāt).*

structure, wherein the Throne and the Footstool are necessarily motionless instruments for the determination of different modalities and durations of time.

Ibn ʿArabī's paradisaical domain consists of eight Gardens, seven of which are hierarchically ordered in seven levels, and an eighth superior one, al-Wasīla, cutting across all the levels assigned to the Prophet Muḥammad.[74] The seven Gardens, in a descending order, are ʿAdan, al-Firdaws, al-Naʿīm, al-Ma'wā, al-Khuld, Dār al-Salām, and Dār al-Maqāma. The names derive from various Quranic verses, and the structure reflects the order of the divine presence: the seven Gardens correspond to the seven principal names, and al-Wasīla represents the Essence. As the Essence supports all the names, al-Wasīla prevails in all the Gardens. The Quran makes numerous references to the Gardens, which Ibn ʿArabī interprets in a layered way.

In addition to the eight-level order, Ibn ʿArabī distinguishes three types of Gardens: Gardens of the Elites *(jannāt al-ikhtiṣāṣ)*, Gardens of Inheritance *(jannāt al-mirāth)*, and Gardens of Deeds *(jannāt al-aʿmāl)*. The first and highest in order refers to the verse: "But God chooses *(yakhtaṣṣ)* for his mercy whom he will" (2:105). The second refers to "Such is the Garden which we cause the devout among our bondmen to inherit" (19:63). The third refers to "Give good tidings unto those who believe and do good deeds, that theirs are Gardens underneath which rivers flow" (2:25). In this hierarchy Ibn ʿArabī focuses on what makes people worthy of being chosen, of inheriting and of inhabiting the Gardens. With reference to the repeated imagery of the "Gardens underneath which rivers flow" (85:11), he further elaborates his layout, saying that in every type of Garden God laid out four rivers, so there are twelve rivers in accordance with the order of the astrological signs.[75]

The four rivers represent the main sources of esoteric knowledge.[76] They include a river of unchanging Water *(māʾ ghayr āsin)* representing the science of life *(ʿilm al-ḥayāt)*; a river of Wine *(khamr)* representing the science of the spiritual states *(ʿilm al-aḥwal)*; a river of Honey *(ʿasal)* representing the science of the divine revelation *(ʿilm al-waḥī)* with its many kinds; and a river of Milk *(laban)* representing the science of secrets *(ʿilm al-asrār)*, the kernel of all sciences that God directly reveals to those who devote themselves utterly to him.[77] Tied directly to the tripartite structure of the human formation *(al-nashʾa al-insāniyya)*—sensible, spiritual, and imaginary—the fourfold pattern of sciences generates twelve different types of knowledge in accordance with the duodenary structure of the zodiac. Ibn ʿArabī explains:

> These are four sciences, while man's formation is threefold: an inward, ideal, spiritual formation; an outward, sensible, natural formation; and an intermediary, isthmian-bodily, imaginal formation. Through each formation man has a distinct share in each of the four rivers, with each share having an independent river, the taste of which differs according to the formation. What man perceives of a river by the senses is other than what he conceives of it by imagination, and what he conceives by imagination is other than the meaning he intellectualizes. This is the order of every formation. So for man there are twelve rivers: four in the Garden of the Elites, four in the Garden of Inheritance, and four in the Garden of Deeds.[78]

Three, 4, and 12 are the numbers that underlie the order of the paradisaical domain. They derive from the duodenary structure of the *aṭlas* sphere, the ruling element. Although the divisions of the *aṭlas* sphere are twelve, Ibn ʿArabī explains, they are of four different natures: aqueous, terrestrial, aerial, and igneous. The quadrature derives from the four pillars of the Throne. Nature rules over all modalities of being in the world of creation, including that of the Gardens, through the agency of the *aṭlas* sphere. Ibn ʿArabī identifies three modalities of being, *manāzil* (dwellings)—the present world *(dunyā)*, the intermediary world (*barzakh*, of the dead awaiting the second creation), and the future world *(ākhira)*—with each having distinct existential conditions. Different though they may be, these three worlds are nevertheless subject to the ruling of Nature, with

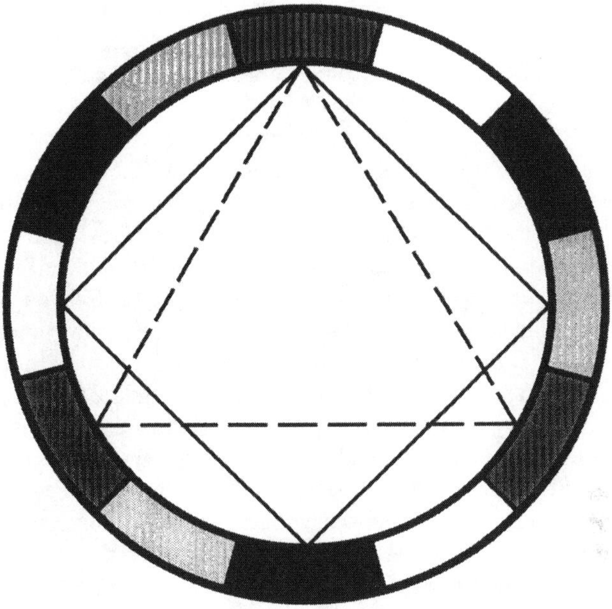

Fig. 3.7 The duodenary structure of the *aṭlas* sphere according to Ibn ʿArabī.

each requiring four distinct signs to mediate the natural processes in ways unique to its modality of being; hence, the twelve signs of the constellation *(burūj)*.[79] The triple repetition of the quadrature results in the signs that are related by the triangle being of the same nature and tendency, whereas the signs that are related by the square being of different nature and tendency.

Since the *aṭlas* sphere is a composed cosmic entity, Ibn ʿArabī says, Nature rules over it through the elements, fire, air, water, and earth, and not through the simple tendencies of heat, cold, dryness, and moistness. Thus Nature differentiates the signs into Igneous (hot and dry), Aerial (hot and wet), Aqueous (dry and wet), and Terrestrial (cold and dry). In their turn, the signs generate in the corporeal domain the spheres of the natural elements (fire, air, water, and earth) whereby generation and corruption occur. Their generative pattern is structured in the following order:

1. Igneous: Aries, Leo, and Sagittarius are the generators of the sphere of fire.
2. Aerial: Gemini, Libra, and Aquarius are the generators of the sphere of air.

3. Aqueous: Cancer, Scorpio, and Pisces are the generators of the sphere of water.
4. Terrestrial: Taurus, Virgo, and Capricorn are the generators of the sphere of earth.

The structure of the *aṭlas* sphere corresponds to the higher cosmic forms within which it is contained, while determining the structure of the Gardens which itself contains. The divine order of universal manifestation projects once more into the domain of the celestial Gardens, manifesting itself in a new cosmic modality. Quadrature is revealed in the fourfold division of the *aṭlas* sphere, in the four natural elements generated by the zodiac, and in the four rivers of esoteric knowledge. Triplicity is revealed in the threefold division of the *aṭlas* sphere, in the three modalities of being, and in the three levels of Gardens. The duodenary pattern of the sphere of the constellations is yet another cosmic manifestation of the original productive marriage of triplicity and quadrature. Centrality and axiality, however, are revealed in the tree of *ṭūbā*, which stands at the center of the Gardens. The tree represents Universal Man, designating, as it were, his place in the Gardens. It relates to the rest of the trees in the Gardens as Adam relates to humankind. God planted it with his own hand in the same way he created Adam. He also breathed the spirit into it, rendering it the most splendid of all trees. It rises above the fence of the Garden of Eden, where God planted it, and its branches spread over other Gardens. Its roots are in the soil of our world and its fruits in paradise.

Heaven and Earth

Within the sphere of the fixed stars, the ground of the Gardens, God unfolded the heavens and the earths, the world of space and time as we know it. Ibn ʿArabī's diagram of this world shows the sphere of the fixed stars with the twenty-eight mansions of the moon (*manāzil al-qamar* (36:38)), the seven domes of the heavens resting upon their respective layers of the earths *(al-arḍūn)*, the four kingdoms, and the Universal Man. At the center of the diagram a vertical line, identified as *ʿamad* (pillars), is shown, representing the *invisible* cosmic pillars that hold up the vaults of heaven. Heaven and earth is the last and innermost world in the hierarchy of cosmic manifestation. It is the sensible world of corporeal bodies. The size of this world in relation to the Footstool, a reported tradition says, is as a ring thrown in a desert. Another describes the suspension of heaven and earth in the middle of the Footstool as the suspension of Footstool is in the middle of the Throne, like a lamp hanging down from the sky. In al-Hindī's hierarchy, heaven and earth is the World of Bodies *(ʿālam al-ajsām)*, the sixth state of dense, composed, cosmic entities that is susceptible of

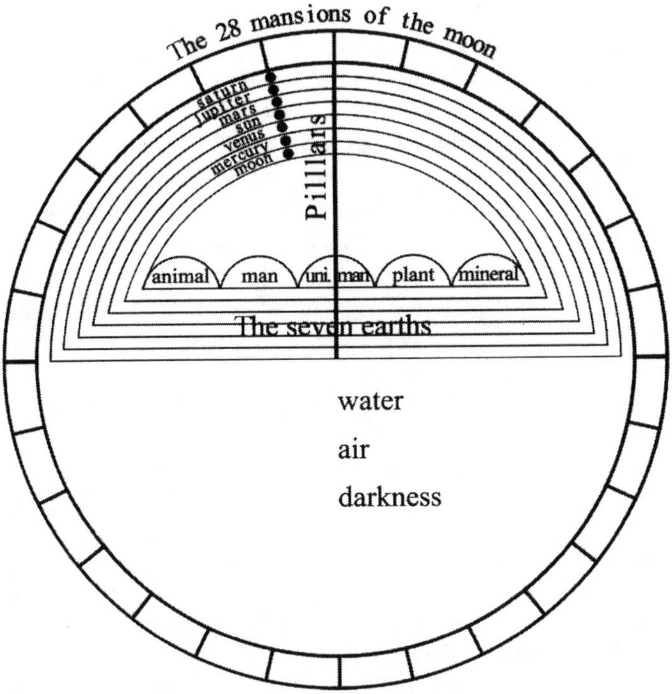

Fig. 3.8 The heavens, the earths, the kingdoms, and Universal Man as invisible support according to Ibn ʿArabī *(Futūḥāt)*.

division, portioning, separation, and conjunction. It is the necessary foundation for man to whom al-Hindī designates the seventh and final state of Being.

The form and the structure of the corporeal world are elaborated in many Quranic references, prophetic sayings, and rich folkloric tradition. According to the Quran, "God it is who has created seven heavens, and of the earth the like thereof" (65:12). Upon the flat expanse of the earths, the Prophet is reported to have said, the skies are constructed "like a dome," an image that seems to derive from the immediate spatial experience.[80] The Quranic descriptions of heaven and earth tend to support such interpretation: "Who has appointed the earth a resting-place *(firāsh)* for you, and the sky a canopy (*binā*ʾ)" (2:22). *Firāsh*, from *farasha*, literally "to extend," "to spread out," and "to furnish," gives the meaning of furnishing the earth by spreading it out in order to accommodate human existence. *Bināʾ*, from *banā*, "to build," "to construct," is often interpreted as "a roof upon the earth in the form of a dome."[81] The Prophet, further elaborating this image, is reported to have said that God created the earths *(al-arḍūn)* flat, seven in number, laid successively one below the other, and each being smaller than the one below it. The seven skies *(al-samāwāt)*, similarly laid one above the other, are domical in shape, and each is smaller than the one above it. Each sky rests on the extremities of its respective layer of earth, thus forming a series of domes placed within each other and separated by a distance of five hundred years of travel. Ibn ʿArabī reproduces these descriptions in a graphic form, illustrating the way in which medieval Muslims conceived of the cosmic structure of heavens and earths. Commenting on his diagram, Ibn ʿArabī reiterates the Prophet's descriptions: "God made the seven skies resting upon earth like domes. Upon each earth, which is spread out like a carpet, a sky like a hemisphere rests on its ends. God spread out the earth so that the sky could stand upon it."[82]

Early Islamic narratives describe the earth as being spread out on the back of a fish, a whale *(ḥūt)* called "*nūn*" (literally "N"), whose ends touch the extremities of the sky.[83] The meaning of this description may be understood with reference to the symbolism of the Arabic letter *nūn*. Geometrically, the letter *nūn* is written as half a circle with a diacritical point representing its center. Originally, Ibn ʿArabī says, the *nūn* was a complete circle, representing the spherical form of the world. But since the world is divided into two equivalent halves—the sensible and the intelligible—the letter *nūn* is likewise divided into two corresponding halves—inscribed and implied or visible and invisible. The analogy is inverted as a mirror image, however. The inscribed lower half of the *nūn* represents the visible upper part of the world, whereas the implied upper half of the *nūn* represents the invisible lower part of the world.

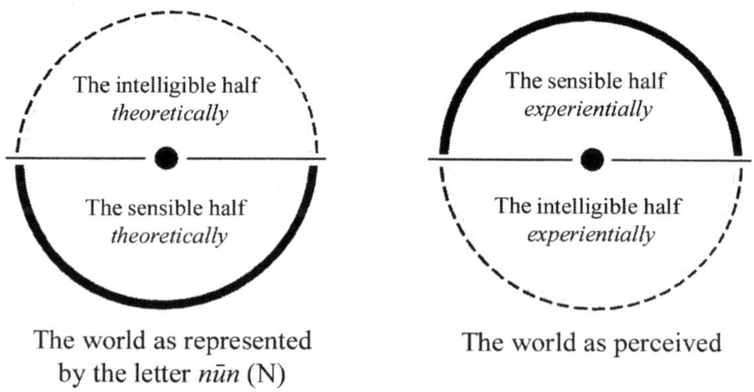

Fig. 3.9 The formal correspondence between the letter *nūn* (N) and the world according to Ibn ʿArabī.

The directional differentiation of *lower* and *upper* has an experiential as well as a symbolic reference. Experientially, standing on the flat expanse of the earth under the hemispherical dome of the sky the invisible half of the sphere is always below us. Symbolically, however, we normally refer to the invisible world of *higher* realities, the world of the unseen, in contrast to the *lower* visible world of shadows, the world of the seen. The inversion here reminds us of the way in which the manifest triplicity and hidden quadrature projects through syllogistical reasoning into the sensible world in an inverse manner—hidden triplicity and manifest quadrature.[84] Acting as a mirror, the letter *nūn* depicts the form of heaven and earth in an inverse manner. The spreading out of the earth on the back of *nūn*, the fish, whose ends touch the ends of the sky, may then be seen as an extension of the flat layers of earth between the two ends of the letter *nūn*. The earth thus becomes the diameter of the circle of *nūn*, the center of which is the original rock.[85]

Early cosmological narratives describe a central rock *(ṣakhra)*, upon which the fish rests, as standing below the lowest earth. This rock is thought of as the foundation upon which the bearers of the Throne stand. A tradition says: "The rock which is beneath the earth is the end of the created world; upon its borders there are four angels, whose heads are below the Throne."[86] At one time, the narrative says, the fish moved so the earth swayed and became unstable and uninhabitable. In order to stabilize it, God "cast into the earth firm mountains *(rawāsī)*" (31:10), the largest of which is mount *qāf* (literally, "Q"), which is described as encompassing the earth, as the perimeter to which the sky adheres, and as the source from which the vault of heaven derives its green color. It is also described as being connected to the rock by means of veins or roots that hold the whole earth firm. When God wants to quake a spot on earth he simply moves the root to which this spot is attached.[87]

In *ʿUqlat al-Mustawfiz*, Ibn ʿArabī describes the process of creating the lower world (*al-dunyā*, from *danā*, to "draw nearer," as opposed to *al-ākhirā*, the "future world," from *akhkhara*, "to delay"). After creating the Intellect, the Soul, the Throne, the Footstool, the *aṭlas* sphere, and the sphere of the fixed stars, he says, the divine gaze *(al-naẓar)* and willed orientation *(al-tawajjuh al-irādī)* were directed toward the creation of the sensible world. First, God ordered an angel to descend in the depth of the space to its innermost point to form the center. This center was to the world what the sacrum *(ʿajb al-dhanab)* was to the animal body: the birthplace and the foundation of its formation. "It is the part that does not decompose *(lā yablā)*," Ibn ʿArabī adds, "the place of attention of the supreme element *(al-ʿunṣur al-aʿẓam)*, from whose brief attention *(iltifāta)* the Intellect is created." As directed, the angel descended to the center of the world and positioned the rock, reaffirming the divine order of things wherein the center is always the source from which things proceed forth and to which they will eventually return. The angel rotated the earth's sphere, making "that which surrounds the center an immense spherical rock, and in the center of that rock the angel placed an animal with a green leaf in its mouth."[88]

This view introduces an interesting shift in the creative process. Up to this stage, God followed a linear descending order. After creating the sphere of the fixed stars, however, God changed procedure by first founding the central *rock* and then unfolding the heavens and earths in an ascending order. Contrary to what many ancient philosophers had believed, Ibn ʿArabī argues, the first thing laid in the corporeal world was the center, around which the skies were then constructed. The earth, as a center, was first laid out and then the skies were formed following the three-dimensional structure of the human formation.[89] The rock formed the "sacrum" of the world's structure, the focal point whence the body of the world unfolded in the six directions—front and back, left and right, up and down—materializing the spatial structure of the human presence.

Ibn ʿArabī's view of the creation and structure of the corporeal world can be traced in the writings of various scholars. For example, al-Zamakhsharī (d. 1144) writes: "The creation of the substance of the earth is anterior to the creation of heaven; but the spreading out of the earth is posterior to it."[90] Al-Diyārbakrī (d. 1582) presents a similar view: "When God began to create things, He created the *turbah* before heaven; when He had created the heavens and divided them into seven stages, He spread out the earth."[91] A popular tradition also relates that "Allah created a *jawharah*, a substance; thereupon he contemplated it with a majestic gaze, so that it melted; then a vapour rose from it, which gave origin to the sky; thereupon the earth was created from the remainder of the *jawharah*."[92] The "green leaf" in the mouth of the mythical animal that dwells at the rock signifies the source of life, immortality, and divine knowledge. In his *Tafsīr*, Ibn Kathīr says that "in the center of this rock there is a spring called "Life" *(ḥayāt)* whose water renders alive whatever it reaches."[93] The rock is also associated with the legendary Quranic figure al-Khiḍr, unto whom God says: "We had given mercy from us, and had taught him knowledge from our presence" (18:66). Al-Khiḍr, whose name connotes the idea of "green," is he "who attained the source of life, has drunk of the water of immortality, and consequently knows neither old age nor death. He is the 'Eternal Youth.'"[94]

These metaphorical expressions show the significance of centrality in the cosmological scheme. Axiality is equally significant. A tradition says that God has a pillar of light *(ʿamūd al-nūr)*, whose base is below the seventh earth and whose top is below the Throne. The pillar vibrates whenever one testifies to God's unity.[95] This pillar of light connects heaven and earth, acting as a channel of communication that is exteriorized through the pillar's vibration. This pillar echoes the concept of Muḥammad as a column of light, already discussed. Penetrating the seven layers of the earth and the seven vaults of heaven, this pillar acts as *axis mundi*, around which existence revolves. It is a direct spatial expression of the axiality of the human presence. When Adam was brought down from paradise, a tradition tells, he was so tall that his head was in heaven, and his feet were on the earth. He could even hear the angels glorifying God.[96] In his diagram of the corporeal world, Ibn ʿArabī depicts the form of the Universal Man by the form of the *axis mundi*—*al-ʿamad* that "you cannot see" (13:2)—without which the vault of heaven would collapse.[97]

Al-insān al-kāmil, the cosmic pillar, is also referred to as "*al-quṭb*," the pole around which the world revolves. The Arabic term *quṭb* derives from the trilateral root *q.ṭ.b.*, "to bring together." As the epitome of all manifested realities, Universal Man *brings together* in his being all possibilities and all modes of existence. He expands both horizontally and vertically: his horizontal expansion is his realization of all cosmic realities, while his vertical expansion is his realization of divine realities. His horizontal expansion is fourfold in that it comprises the realities of the four kingdoms—the mineral, plant, animal, and human—each of which represents certain modes of existence and actualizes a particular ensemble of possibilities. His vertical expansion traces a return passage from quadrature to unity. The idea of Universal Man recapitulates both cosmic and divine realities, and the notion of axiality ensures continuous communication between the higher and the lower worlds and harmony between Man's parental domains: his celestial fathers and terrestrial mothers.[98]

The term *quṭb* is the name for the central piece around which a millstone rotates. Acting as a motionless hub for the rotating upper part, the *quṭb* is firmly fixed in the still lower part of the millstone, hence the imagery of *axis mundi*. This is taken to represent the firm earthly rootedness of Universal Man, while his verticality acts as the unchanging pole around which the wheel of change revolves. "A millstone rotates only about its pole *(quṭb)* when this pole is in the millstone," Ibn ʿArabī writes, "for the pole is its firm essence, which is not susceptible of movement nor transposition during the state of rotation."[99]

The revolution of the stars and heavenly bodies in the corporeal domain is also seen as following the metaphysical order of things in many and different ways. The spatial journey of the sun around the stationary earth, for example, engenders space and time, revealing the intertwined relationship between triplicity and quadrature. The annual journey of the sun is punctuated by four nodal points—two solstices and two equinoxes—dividing the ecliptic into four qualitatively distinct intervals and marking out the cardinal directions of space. The cyclical rotation of the moon around the earth, which regulates time in the Islamic calendar, also has quadrature as its basic pattern. The twenty-eight phases of the lunar cycle involve four recurrent intervals, each with seven phases (4 × 7). No moon *(al-maḥāq)* and full moon *(al-badr)* mark the beginning and the middle of the cycle, the two extremes analogous to the solstices, whereas the half moons *(al-tarbīʿ)* mark two intermediate points similar to the equinoxes. The seven phases of every interval, which determine the week, are in themselves generated numerically

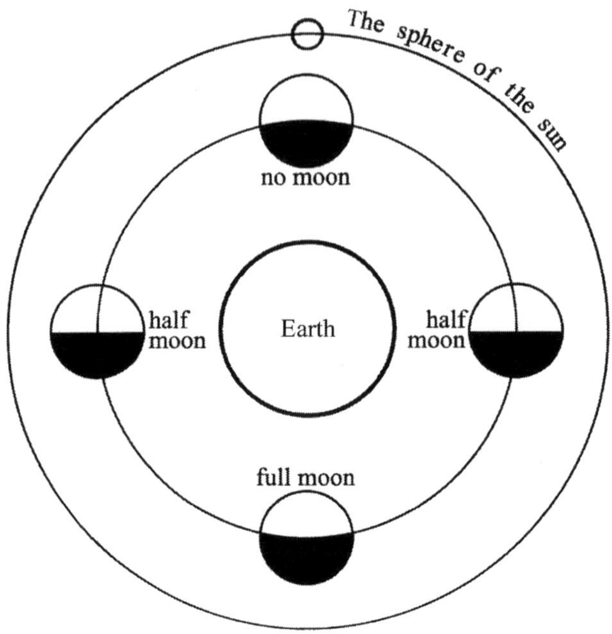

Fig. 3.10 The four nodal points of the moon's monthly cycle according to al-ʿĀmilī ("Tashrīḥ," MS. 3103).

by three and four (3+4=7), unfolding yet another cosmic modality of the primordial bond between triplicity and quadrature.

Within the domain of heaven and earth the metaphysical order of things is revealed in many forms. Centrality is revealed in the rock, the sacrum *(ʿajb al-dhanab)* of the world's body, and the primordial mountain. Circularity is revealed in the form of heavenly bodies and their orbits; in the domical form of the skies; in the form of *nūn*, the fish; and in mount Qāf that encompasses the earth. The pattern of proliferation into a multitude of secondary centers is revealed in the ubiquitous presence of mount Qāf, symbolized by the links it has to all mountains and every place on earth. "There is no one country amongst all countries, nor a city amongst all cities, nor a town amongst all towns but has a root of its roots," a tradition affirms. Another adds, "nor is there any mountain of all mountains but has a root in Qāf."[100] Axiality is revealed in the pillar of light, in the primordial mountain, and in Universal Man, the cosmic pillar. Quadrature is revealed in the four angels standing on the rock, the four kingdoms, the four nodal points of the sun, the four directions of space, and the four intervals of the moon. Triplicity and quadrature are also embodied in the corporeal conditions of space and time.

Space and Time

Perhaps nowhere the presence of the metaphysical order of triplicity and quadrature is more immediate and tangible to us than in our existential conditions of space and time. The three dimensions of space—length, breadth, and depth—spatialize triplicity, whereas the four divisions of time—day, week, month, and year—temporalize quadrature. Immanent yet elusive, tangible yet hard to define, space and time have their roots deep in the infinity and eternity of divinity. Throughout history and across all traditions, the human mind has marveled at the nature of these bounding conditions while constantly searching for ways to understand them. Today, science has taken the lead, but in premodern times it was the religio-philosophical imagination that provided the answers.

Following Judaism and Christianity, Islam reaffirmed the narrative of the creation: "Your lord is God who created the heavens and the earth in six days" (7:54). The six days of the creation presents a complex paradox and raises some interesting questions. If the existence of space and time is evidently tied to the existence of heaven and earth and the movement of the heavenly bodies, how could heaven and earth be created in six days? If the "day" mentioned in this verse refers to the temporal duration that we experience between two successive risings of the sun, no days should have existed before the existence of the stars and planets, and the six days of the creation remain inexplicable in terms of our time. This has often been resolved either by maintaining that the six days of the creation are presumed durations or by differentiating between two spatio-temporal modalities: divine and human. In Islamic cosmology, this was achieved by distinguishing the spatio-temporal modality of the Throne and Footstool from that of heaven and earth. Through his intricate structure and a double movement scheme, Ibn ʿArabī was able to maintain that space and time already existed before the creation of heaven and earth, and that a day as a measuring unit was also differentiated in time by reference to the position of the divine Feet on the Footstool. This is a divine day, of course, as in "a day with God is as a thousand years of what you reckon" (22:47). Unlike our time, however, the divine time has neither daytimes *(nahār)*, nor nighttimes *(layl)*, nor weeks, nor months, nor years, nor seasons, for all of these relate to the sun and the moon.[101] There is, instead, pure duration of only one ever-recurring day.

As already discussed, the manifestation of space and time in their pure form first occurred in the Universal Matter, which was specially conditioned for this purpose. Their presence in the world of creation coincided with the production of the Absolute Body *(al-jism al-kull)*, whose existence was mediated by three cosmic agents: the Intellect, the Soul, and the vacuum *(al-khalāʾ)*. God first brought Universal Matter into existence, Ibn ʿArabī says, within which he then unfolded the form of the three dimensions. The length, he says, was from the Intellect, the breadth from the Soul, and the depth was the vacuum, extending from the outermost perimeter to the innermost center.[102] God made the Universal Body circular in form, Ibn ʿArabī adds, filling up with it the entire vacuum—the imaginary extension without a body—and leaving outside it neither vacuum nor plenum.[103] In this sense, the Universal Body stands for the first spatialization of both the divine and the human presences and the materialization of the geometry of being.

Nature rules over the Universal Body, conditioning its possibilities, Ibn ʿArabī explains, while being the principal generator of time. In our spatio-temporal modality, nature first unfolded the annual measure, the year, and differentiated the four seasons—spring, summer, autumn, and winter—whose quadrature was then reflected in the four divisions of time—year, month, week, and day. The quadrature of the seasons was manifested by the sun traversing the signs of the constellations, which were also divided by nature into Igneous, Aerial, Aqueous, and Terrestrial, according to the division of the *arkān* into fire, air, water, and earth, which in turn reflected the divine creative quadrature.[104] Movement *(ḥaraka)* was the main principle of generation brought about by the concurrence of space and time. Thought of as a form imposed on the Body by the Universal Soul, motion directly linked space and time to the generation of life.[105] "Time is associated with the motion of the Body," the Ikhwān explain, "and the Body is passively generated by the Soul. As the Soul made the Universal Body spherical in shape, which is the noblest of all shapes, it also made its motion circular, which is the noblest of all motions."[106]

Ibn ʿArabī reflects on the reasons behind the original motion and its circular form. The initial movement, he says, was caused by disequilibrium in the contrasting natural forces (heat-cold, moistness-dryness). In their original state of equilibrium nothing occurs: there can be neither bringing-into-existence nor production. The equilibrium was broken when heat dominated other forces, and as the amount of heat in the body of the sphere increased, the sphere moved. But there was nowhere to move to, for it had already filled up the entire vacuum. Under the force of disequilibrium, it moved in its place about its center, a movement of the middle.[107] This is best represented by the motion of a millstone: while every part is moving from one place to another by the movement of the whole, the whole itself does not change its location by the movement of the parts. This is the case with every circular motion, Ibn ʿArabī says, "it is moving–still." With regard to the whole, it does not evacuate its space by translocation, yet its parts remain in motion.[108]

The Universal Body is an imaginary entity, a necessary conceptual foundation for all three-dimensional bodies in the world of creation. Its first materialization is the Throne. As mentioned earlier, some astronomers identify the Throne with the *aṭlas* sphere or the sphere of the constellations, the motion of which then becomes the first generator of time. Ibn ʿArabī, however, places the *aṭlas* within the Throne in order to differentiate three modalities of time: divine, paradisaical, and human. As the cosmic progenitor of space and time, the *aṭlas* sphere plays a mediatory role, whose motion can be determined both from within and from without. Being without stars, the motion of the *aṭlas* engenders a pure, undifferentiated duration. On its own it lacks any distinguishable reference point that may differentiate its incessant movement into recurrent cycles. Viewed from within, with reference to the sphere with fixed stars and the movement of planets and the stars, its pure motion becomes differentiate into the human temporal durations (day, week, month, year). Seen from within but without reference to the sphere with fixed stars and planetary skies, we have one pure paradisaical duration. Seen from without, however, with references to the position of the divine Feet on the Footstool we have the divine durations, to which God refers in the six days of the creation.[109]

As is the order of being, the divine modality of time forms the model for the human temporality. In articulating the relationship between the divine and human temporality, Ibn ʿArabī first distinguishes the movements of the stars from the motion of the *aṭlas* sphere itself. The revolution of the stars within the *aṭlas* sphere, he posits, is subject to two types of movements: natural *(ṭabīʿiyya)*

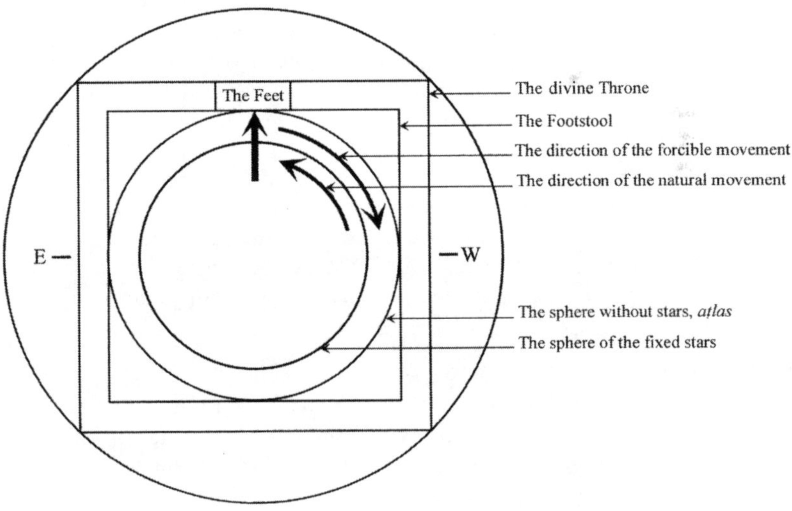

Fig. 3.11 Differentiating the motion of the *aṭlas* sphere by reference to divine Feet according to Ibn ʿArabī.

and forcible *(qasriyya)*.[110] The natural movement is that whereby the stars traverse the *aṭlas* sphere from west to east, as seen by the eye, whereas the forcible movement is the reverse one whereby the stars move with the movement of the *aṭlas* sphere from east to west.[111] The situation is like an ant on a piece of material that is being pulled westward while the ant is moving eastward: the ant is simultaneously moving to the east and to the west. This is the situation of the stars, he says: at the same instant in which they are naturally moving from west to east, they are also forcibly moving from east to west. Humans work with the natural motion, while God works with the forcible. By means of the position of his Feet on the Footstool, God differentiates the primordial motion of the *aṭlas* sphere into seven recurrent cycles, or days, in accordance with the seven principal attributes of the divine presence. Ibn ʿArabī explains:

> By the existence of the *aṭlas* sphere there occurred the seven days, the months, and the years. But these times were not determined until after God created inside this sphere the signs whereby these times were distinguished. The only duration this sphere determines is the day, which is one cycle determined by the position of the Foot on the Footstool. So it is determined from above, and the measure of one complete cycle is called "day" *(yawm)*. Because of the homogeneity of this sphere, this day is known only to God most high. The starting point of its movement coincides with the first degree of Gemini, which is among the Aerial signs, facing the Foot. The first day manifested in the world was in the first degree of Gemini, and that day was called "al-Aḥad" (Sunday) . . . Upon the end of the first cycle the sphere started another movement . . . This second movement was called "al-Ithnayn" (Monday), and so on until the seven cyclical movements were completed, one divine attribute determining each movement. As the [principal] attributes are no more than seven, the days of the aeon *(al-dahr)* cannot be more than seven, not even by a day, for there is nothing that may necessitate it. Thus, the ruling returned to the first attribute that rotated the sphere [again] and the name Aḥad became associated with it . . . [For the new cycle,] however, it was more appropriate to be the eighth with regard to the cycles, but since its existence was caused by the same first attribute its name did not change. Similarly, the cycle that followed, and so was the following one until the seven cycles [were once again completed].[112]

Accordingly, each divine attribute is seen to engender one entire cycle or day, during which period this attribute dominates over other attributes, causing its characteristic to inhere in all creatures. During the cycle which is engendered and dominated by the divine attribute of Hearing, for example, all creatures receive the characteristic of hearing by virtue of which they become able to hear. The same goes for the rest of the principal attributes. Since the creative process begins with the utterance of the primordial word, *hearing* was the first characteristic the creatures received so that they became able to respond to the divine creative command "Be!" According to Ibn ʿArabī, the creative process took place in the following order:

Hearing	Sunday	al-Aḥad
Living	Monday	al-Ithnayn
Seeing	Tuesday	al-Thulathāʾ
Will	Wednesday	al-Arbiʿāʾ
Power	Thursday	al-Khamīs
Knowledge	Friday	al-Jumuʿa
Speech	Saturday	al-Sabt[113]

Representing the governing principles of universal manifestation, the seven divine attributes at once determine the cyclical revolution of the *aṭlas* sphere and necessitate the creation of seven planets. This ensured the harmony and continuity between the divine and human modalities of time. In an interesting treatise on time and eternity, al-Qāshānī says that the seven principal divine names required intermediaries to ensure the continuity of their domination over all things in the corporeal world. So these names caused the existence of the seven revolving planets together with their spheres and made them the presidents and chieftains for directing the affairs of the present world. Thus the seven planets embody the dominating power of the seven divine names, representing the cosmic intermediaries between the immutable world of archetypes and the earthly creatures. The movements of the seven planets differentiate the pure temporal duration of the first motion into recurrent cycles of measured duration in accordance with the divine model. Al-Qāshānī explains:

> If you consider the first motion and the extension of its duration, which is undifferentiated time *(al-zamān al-muṭlaq)*, disregarding what is below it, it has neither beginning nor end nor division. But if you relate the sun to a particular point, any point whatsoever, the year, whose every cycle is the return of the sun to that point, begins by the movement of the sun whereby it traverses the parts of the sphere of the constellations. By this movement the [pure] duration is articulated into years; the year, in regard to the sun's traversing of the constellations, into months; the months into weeks; the weeks, in regard to the sun's return to the first point in its diurnal movement, into days; the days into hours; the hours into minutes; the minutes into seconds; then into thirds until now, which is to time as the geometrical point is to a line.[114]

The movements of the seven planets also qualify the pure extension of space. The diurnal and annual journeys of the sun, while differentiating time, simultaneously qualify space by measuring out its extremities and marking out the cardinal directions—east, west, south, north, zenith, and nadir. Through the combined rhythm of their revolutions, the seven planets construct various geometric relationships based on their reciprocal positions against the twelve signs of the zodiac, unfolding as it were the spatial cosmic qualities that are virtually contained in the pure, unqualified space of the *aṭlas* sphere. In Ibn ʿArabī's cosmological scheme, the *aṭlas* sphere with the rock as its center, imagined independently of the planetary skies, gives the image of an all-encompassing sphere that represents space as an undifferentiated totality. Thus imagined, the *aṭlas* provides the cosmic model for the geometric sphere, the spatial expression par excellence of the divine presence. Its empty vastness signifies the divine's all possibility and immutability, while the inexhaustible multitude of its directions signifies the multiplicity of the divine names and attributes.[115] As the inexhaustible multitude of the names is exemplified by seven only, six of which are related to the created world and one, the Living, acting as their principle, so likewise the multiplicity of the directions in space, determined by the lines connecting the point of the center and the points of the sphere's surface, are exemplified by seven only, six of which are on the sphere's surface, determining the six main directions—front and back, left and right, and up and down—and the seventh point, acting as their common principle, the direc-

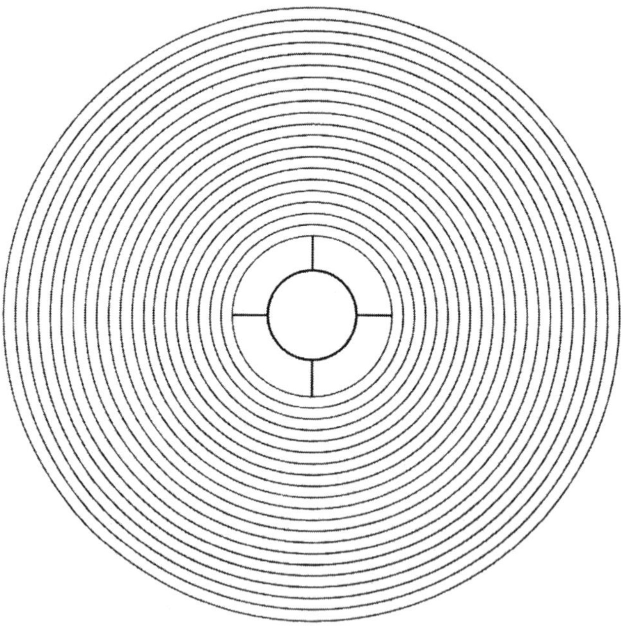

Fig. 3.12 The multilayered world according to Ibn ʿArabī *(Futūḥāt)*.

tionless source. The primordial motion of the *aṭlas* sphere generates a pure duration of time that is, as is the pure extension of space, indefinite, unqualified, and undifferentiated. This mode of time mirrors divine eternity, as time is to us what eternity is to God.[116] Al-Qāshānī even compares the undetermined temporal duration generated by the primordial motion to the inconceivable duration *(imtidād)* of the subsistence of the divine Essence. As the undetermined Essence becomes determined when the seven principal names are related to it, so likewise the undifferentiated motion of the *aṭlas* sphere becomes differentiated when the movements of the seven planets are contrasted with its motion.[117]

In that cosmic manifestation coincides with the utterance of the primordial word, the differentiation of the pure spatio-temporal modality of the *aṭlas* sphere corresponds to the articulation of the primordial sound, the medium through which the primordial word was externalized. The Sufis correlate the twenty-eight letters of the Arabic alphabet, which represent as many articulations of an unobstructed breath emanating from the heart, with the twenty-eight mansions of the moon. The articulated sounds of the letters are "the microcosmic and human expression of the essential determinations of the divine Breath, which is itself the prime motivation of the cosmic cycles."[118] Ibn ʿArabī says that contrary to what people think that "the mansions of the moon represent the models of the letters; it is the twenty-eight sounds which determine the lunar mansions."[119] The twenty-eight mansions of the moon also correspond to twenty-eight divine names. The process of the sonorous differentiation proceeds in a successive order from the letter *alif* (A) to the letter *wāw* (W). As the *alif* does not admit any of the vocalizing motions *(ḥarakāt)*, it is unpronounceable. It is therefore represented by the *hamza* (hiatus), which "is not properly speaking a sound, but only a transitory instance between silence and locution."[120] The *hamza*, the threshold between the silence of nonexistence and the sound of existence, corresponds to the Universal Intellect and coincides with the spring equinox. The *hamza* represents the unpronounceable *alif*, which is to the letters what the one is to the numbers. In the same way that one is not a number but the principle of numbers, the *alif* is not a sound but the principle of sounds. The first articulated sound is *hāʾ* (H). Proceeding from the innermost source, the heart, *hāʾ* is the least differentiated and articulated consonant; it is the audible reflection of the inaudible *alif*. The articulation of sounds proceeds from *hāʾ*, passing through the guttural, palatal, and dental consonants, to the last and outermost labial consonant *wāw*. The first and the last consonants in this process form the Arabic word *huwa*, "he," which sums up the whole order of being: "He is the first and the last, and the outward and the inward; and he is knower of all things" (57:3).

Chapter 4

Architectural Order

Gazing at the Sky

Al-Ghazālī's analogy that served as the starting point of this excursion into premodern Islamic cosmology and metaphysics was concerned with the procedural aspects of the creation, with how God and an architect alike first produce a written or drawn *exemplar* in accordance with which an *object* is then brought into existence. The analogy does not tell us much about the contents of these exemplars, nor about whether and how the architect interprets the divine paradigms and casts them into architectural forms. Even in his more detailed discussion of the divine names, al-Ghazālī's main concern remains the process and protocol of production. In a treatise on the wisdom of God's creatures, al-Ghazālī gives us other clues. With references to the verses: "Have they not then observed the sky above them, how we have constructed it and beautified it, and how there are no rifts therein?" (50:6) and "God it is who has created seven heavens, and of the earth the like thereof" (65:12) he writes:

> Know, may God treat you with mercy, that if you reflect in your mind upon this world you will find it like a built house equipped with everything one needs. The sky is raised as a roof, the ground is stretched out as a carpet, the stars are hung like lamps, and the substances are stored as treasures. Everything is prepared and specifically formed for a purpose. Man acts as the owner of the house who is in charge of its contents. The varieties of plants are designated for his needs, and the species of animals are dedicated to his interests. God also created heaven and made its color most appropriate and strengthening for his vision. For if it was pure rays and lights it would have harmed the onlooker. Looking at the green and blue, however, is suitable for the human sight, as the

> souls find felicity and comfort in gazing upon the vastness of the sky, and especially when the stars are shining and moonlight is clear. For this reason, the kings adorn the ceilings of their courts with patterns and decorations that give the viewer comfort and delight. Yet, as the viewer continues to look at this adornment he becomes bored with it and loses what he used to find in his visual experience of felicity and delight. This is unlike gazing upon the heaven and its adornment, to which those displeased by whatever reason—be they kings or lay people—turn their sight seeking delight both in the sky and the vastness of space. As the wise men say: "you will have of comfort and delight in your house just as much as you have of the sky."[1]

That the cosmos, in its complexity, beauty, and order, conceals a profound divine wisdom is beyond question for most Muslims. It is a core theme in the Quran that enjoins Muslims to reflect upon the wonders, beauty, and wisdom of God's creation in order to deduce lessons, guidance, and meanings for their worldly practices. A whole genre of literature concerned with the wisdom of God's creations proliferated in premodern Islam. Even Islamic historiography was predicated on the notion of *iʿtibār*, the need to reflect and "take lessons" from the events of past generations in order to understand God's hidden wisdom. The search for and praise of divine wisdom is traceable in a wide range of Islamic literature, and especially in the later Ottoman writings on architecture. The seventeenth-century text of *Risāle-i Miʿmāriyye* wonders about the architecture of the world:

> What is this exalted mosque and retreat for witnessing?
> What is this lofty vault and lamp ornament?
> What is this bright window, what is this luminous taper?
> What is this wonderful creation, and what is this beauteous form?
> What is this vault of heaven, and what is this surface of the world?
> What is this lofty arch, and what is this great pavilion?
> What is this? Who made such an edifice?
> Without drawings and without mathematics and without analogy?[2]

Such reflections, while being concerned with procedure and aesthetics, support the assumption of a deeper connection between the divine act of creating and the human act of designing. In the preceding, I have explored the "contents" of the divine exemplar from the mystical perspective, focusing on what the Sufis consider to be the consistent, underlying *niẓām* (order) of the universe, the thread that ties together all divine, cosmic, and human manifestations. So far, my focus has been on the divine side of the analogy; here I will turn to the architect's side to discuss the tectonic embodiment of the universal order in architecture.

Ordering Spaces

A simple examination of a range of surviving premodern Islamic buildings reveals a discernible preference for geometrically ordered spaces with isotropic spatial qualities. There was a tendency to organize spaces symmetrically around a central point and to identify, in one form or another, the cross of directions, regardless of whether or not the cross is aligned with the cardinal points. For example, the formal composition of a dome on a geometrically regular base or a courtyard with a central fountain and four vaulted *īwāns*, doorways, and other symmetrical elements are recurrent in a variety of building types. They were used to form the whole or parts of buildings as diverse as tombs, mosques, palaces, hospitals, caravanserais, public baths, garden pavilions, and schools. This same spatial order is traceable in buildings that serve secular as well as religious purposes. There are rich tectonic and regional variations, of course, but the underlying spatial order remains consistently visible from India to Spain and from the formative period to the late Ottomans. This tendency is not peculiar to Islam though. Notwithstanding the conspicuous differences in tectonic expressions, other traditions reveal a similar sense of ordering. The consistency of this tendency across geo-cultural, temporal, and typological boundaries is taken to suggest a *spatial sensibility* that is more uniquely premodern than Islamic. Shaped by prevailing intellectual and intercultural conditions, as well as by established professional practices, 'spatial sensibility' remains an elusive concept, one whose roots extend well beyond any historically or culturally identifiable boundaries. Yet spatial sensibility is graspable through the sense of ordering and spatial structure it reveals, providing an appropriate focus for broad theoretical analyses of architecture and cosmology across historical periods, building types, and regional variations.

Spatial order, thus identified, is concerned primarily with individual spaces that are pictorially and experientially unified. Large architectural complexes are considered to be composed of a series of interrelated spaces that are pictorially unified. The overall compositional qualities may vary widely, as in the Taj Mahal in Agra, Masjid-i Shāh in Iṣfahān, the Dome of the Rock in Jerusalem, the Sultan Ḥasan school in Cairo, and the courtyard gardens of Alhambra. However, the same sense of spatial ordering is clearly visible in all. Focusing on spaces of pictorial unity enables us to appreciate the compositional relationship of the whole and the parts without giving priority to the whole or being concerned about the accretive or accidental nature of compositions that is common in premodern Islamic buildings. It also enables us to correlate elements of different styles and historical periods within an ahistorical frame, such as the early Umayyad Dome of the Rock and the Aqṣa mosque along with the various Mamluk additions, especially the Ashrafiyya school, which are experientially unified in the ensemble of the Ḥaram al-Sharīf in Jerusalem. This approach resonates with the way in which Muslim travelers present their spatial experiences of buildings and landscapes as seen outside their distant and diverse sociohistorical origins. The rich variety of designs and tectonic expressions reveals two types of composition: concentric and linear.

Concentric Composition

The concentric composition represents all architectural designs that are laid out about a stationary center, expressing the spatial order of the three-dimensional cross in a static manner. Two models typify spaces thus ordered: a centralized enclosed space and a centralized open courtyard. A centralized enclosed space includes all architectural spaces that are defined by a geometrically regular base and domical, conical, or other form of centralized roofing. While the roof tends to emphasize unity and centrality, the base tends to emphasize the directionality and spatial deployment. The Umayyad Dome of the Rock and the adjacent four-*īwān* hall of the Mamluk Ashrafiyya school in Jerusalem, recurrently visited described by Muslim travelers, are two immanent examples of the concentric order. Their diverse forms and distant temporality show how a particular spatial sensibility can continue to manifest the same sense of ordering in different forms and across both historical periods and stylistic variations.

Photo 4.1 The Umayyad Dome of the Rock in Jerusalem.

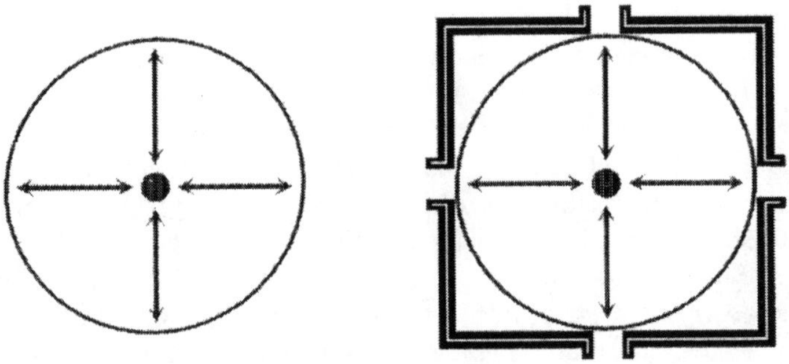

Fig. 4.1 The geometry of the concentric composition.

Fig. 4.2 The formal order of the centralized enclosed space model.

Geometrically, a centralized enclosed space develops from a regular polygon (mostly quadrangular); spatially, it expands from a focal point and evolves symmetrically about a central axis, resulting in a balanced synthesis in all directions. The simplest architectural embodiment of this model is a building with a cubical base and a hemispherical dome. The architectural composition of the Islamic mausoleum, described as "the posterity of the Dome of the Rock," follows this model.[3] Mausoleums can be free-standing structures as in Qubbat al-Ṣulaybiyya, Sāmarrā', and the tomb of the Samanids, Bukhārā, or part of a larger complex, as in the mausoleum of Barqūq and that of Sultan Qayitbay, Cairo. Externally, Islamic mausoleums vary greatly in shape and scale; however, most of them are composed of a geometrically regular base with a domical or conical roof. Internally, they reveal the same centralized order. The cities of the dead in Cairo show the widespread use of this model for tomb architecture. *Gunbads*—mausoleums in tower form—found mostly in Iran, such as the Gunbād-i Qābūs and Gunbād-i Ghāzān Khān, though of dif-

Photo 4.2 The central domed structure at Ibn Tūlūn mosque in Cairo showing the geometry and spatial order of the centralized enclosed space model.

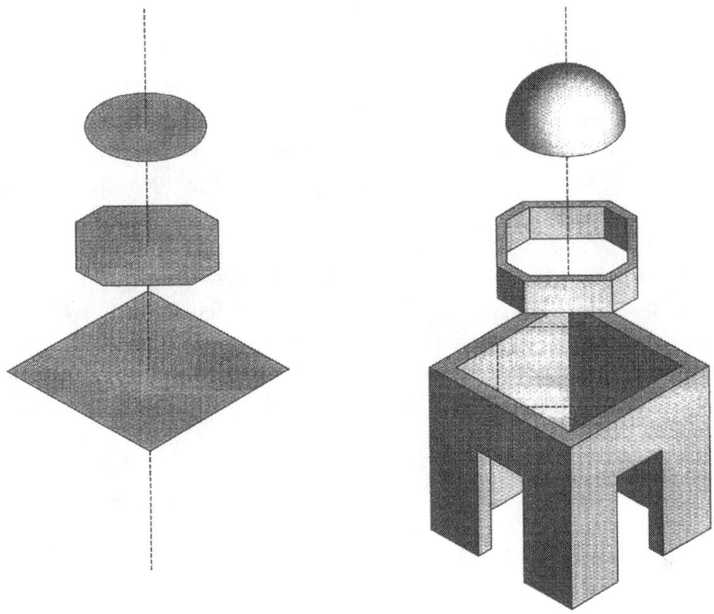

Fig. 4.3 The formal components of the centralized enclosed space of the model.

ferent appearance, are only variations on the theme. They exhibit the same spatial regularity and symmetry but emphasize the vertical axis.[4]

The second form of concentric composition is the centralized open courtyard. It represents all confined, unroofed spaces that are organized symmetrically about a central point. The geometrical order of this composition reveals, like the previous model, the geometrical proliferation of unity into quadrature. The defining surfaces are usually symmetrically articulated in relation to a central axis, leading to a balanced deployment of space in all directions. Far from being merely a negative space, the open courtyard forms an integral and major part of many architectural compositions. In many cases it even forms the major space of which the building is the defining parameter.

The courtyard can form a part of buildings such as mosques, schools, and houses; or on a larger scale it can become a large garden or the central space of an entire city. In the many cases, where the internal courtyard is a predominant spatial element, regular geometrical and spatial qualities are evident. Whether a centralized courtyard is the determining model of an entire architectural complex or only a part of it, it normally expresses, in one form or another, the underlying order of the concentric composition. Many

Islamic gardens follow this model. Domestic courtyard gardens, characterized by a central fountain and regular spatial arrangements can be found throughout the Islamic world.[5] Large gardens, particularly those found in the Iranian plateau and the Indian subcontinent, are particularly expressive of this model. Two complementary spatial arrangements have been identified: gardens inwardly oriented with a major structure, a pavilion or a mausoleum, occupying its center; and gardens outwardly oriented with an empty center and a major structure defining its boundary.[6] Some gardens with a central domed pavilion, such as Hasht Bihesht in Iṣfahān, show a sophisticated synthesis of the centralized enclosed space and centralized open courtyard. The planning of some of the early Islamic cities, such as al-Kūfa, al-Baṣra, and Baghdād also follow the model of a centralized open courtyard. As described by Muslim chroniclers, they were laid out around a large open court *(ṣaḥn)* centered by one or two buildings—a mosque in al-Kūfa and al-Baṣra and a mosque and the caliph's palace in Baghdād, revealing the same underlying spatial order at a larger urban scale.

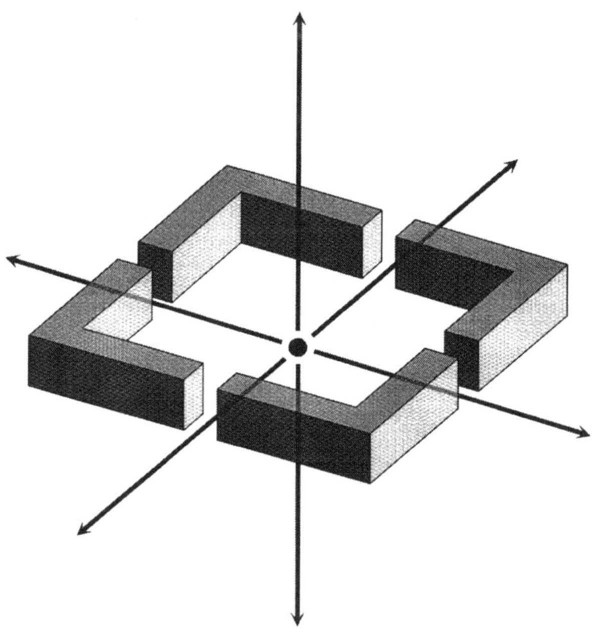

Fig. 4.4 The centralized open courtyard model.

Photo 4.3 The courtyard of the Sultan Ḥasan school in Cairo showing the geometry and spatial order of the centralized open courtyard model.

Linear Composition

The linear composition is a variation on the concentric composition involving repetition. The repetition of a concentrically ordered unit generates a linear composition, conveying motion and extensionality. The linear composition can be seen primarily in premodern bazaars, such as those still existing in Iṣfahān, Kāshān, Aleppo, and Jerusalem. The repetitive form might have been generated by structural necessities, yet the spatial characteristics of the linear spaces are expressive of the same spatial sensibility that underlies the concentric compositions. While maintaining the order of centrality, axiality, and quadrature, the linear composition is created when the stationary center of a concentric space "moves," so to speak, manifesting through this motion a linear space that joins two or more points.[7] In contrast to the concentric composition, the linear composition represents all spaces that are focused by a "moving" center, expressing the underlying spatial order in a dynamic way. Movement enables reiterative exposure to a similar formal unit and spatial structure, the arched base and domed roof, creating a sense of monotony and repetition. Colonnades, porticoes, and spaces covered with a multidomed structure, typical of Ottoman architecture, share with the bazaar its linear, dynamic characteristic. Architecturally, the linear

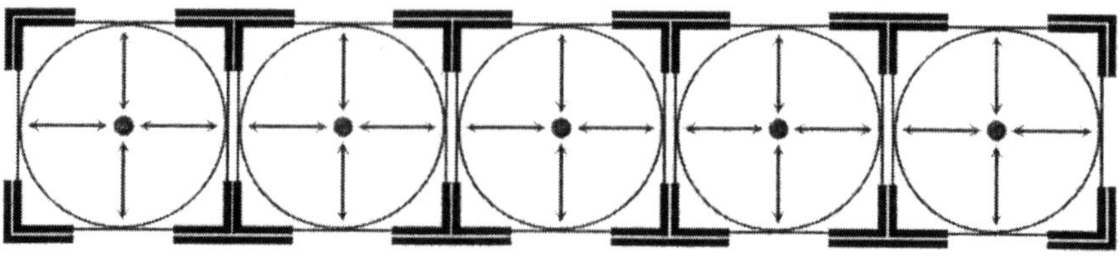

Fig. 4.5 The geometry of the linear composition.

composition is formed by the repetition of a spatial unit, creating a number of individual concentric spaces or "spatial pulses." These units are linked together in a manner analogous to the way beads of a rosary are connected upon its thread. The monotony of linearity is often interrupted when the main route of the bazaar intersects with another or when the entry to a building is emphasized. These interruptions produce a series of nodal points that break the regulating monotony of linearity.

Whereas the static unfolding of space in the concentric composition reveals one center in a pictorially unified space, the dynamic nature of the linear composition manifests a multitude of centers, all of which are of more or less equal importance. As a series of "spatial pulses" they embody in a repetitive manner the same underlying spatial order and reveal similar spatial characteristics. An architectural composition that is concentrically ordered may also comprise a multitude of centers, but usually varying degrees of importance can be distinguished. A geometrical analysis of the plan and form of Taj Mahal, for example, shows how the central space is distinguished in size and articulations from the other similar but smaller spaces, which nonetheless reveal the same underlying spatial order as the whole.[8] From an analogical perspective, one may observe that the concentric composition is the basis from which the linear composition derives, just as the point is thought of as the principle from which the line extends and stillness *(sukūn)* as the state from which motion *(ḥaraka)* proceeds.

The Quest for Principles

Premodern spatial sensibility and modes of spatial ordering were rooted in a complex web of scientific and theological thinking. Geometry, geography, and astronomy, an intertwined set of scientific enterprises, were often under the influence of theological, philosophical, and astrological speculations that were concerned with the origin, order, and purpose of the universe. In the complex matrix of ideas and preoccupations that underpinned premodern modes of thinking, the relentless quest for principles provided a common ground across most disciplines. If existence is meaningful only with God as its principle, then everything else in the world must likewise have a founding principle, including, one would think, the ordering of spaces and making of architecture. This prevailing assumption can be traced in numerous sources. The Ikhwān give an explicit reference. In their treatise on geometry, the Ikhwān speak of the divisions of science introduced by people of antiquity *(al-qudamā')* to teach their children and students. The divisions they identify follow those of the Greeks: mathematics, logic, natural sciences, and metaphysics.[9] Each division, they say, includes many branches, each of which has a principle *(mabda')* that governs its fields and possibilities. Mathematics, for example, includes four sciences: arithmetic, whose principle is the number 1; geometry, whose principle is the point; astronomy, whose principle is the movement of the sun; and music, whose principle is the proportion (the equation of two or more ratios). The same goes for other divisions.

This mode of reasoning constructs, on the one hand, causal links between the principle and its manifestations and, on the other, correspondences between different worlds and modes of being based on the inherent order that binds all principles to their respective worlds. As we have seen, the whole of Sufi cosmology is motivated by a passion to understand how the world unfolded from its primordial principle. This quest for principles is central to our understanding of what shaped the spatial sensibility and sense of ordering of premodern Muslims. From the third century onward, Euclidian geometry and Ptolemaic astronomy prevailed in the Islamic world. As these sciences were themselves underpinned by a quest for principles, it is necessary to understand the scientific frame of geometrical imagining that legitimated Sufi cosmological and cosmogonic interpretations.

Designing necessarily involves geometry, the Ikhwān write, since it involves the manipulation of dimensions and measures according to what the designer conceives of and imagines in their mind.[10] According to Euclidian geometry, everything begins with the point, the principle to which the entire science of geometry, its intelligible and sensible forms, can be reduced. The Ikhwān explain:

> An intelligible line cannot be seen on its own, but only as it lies between two surfaces, like the borderline between sunlight and shade. If there were no sunlight and no shade you would not have seen a line [defined] by two imaginary points. And if you imagine that one of these two points is moving while the other is standing still, until it returns to where the movement began, a plane will occur in your mind. An intelligible surface, too, cannot

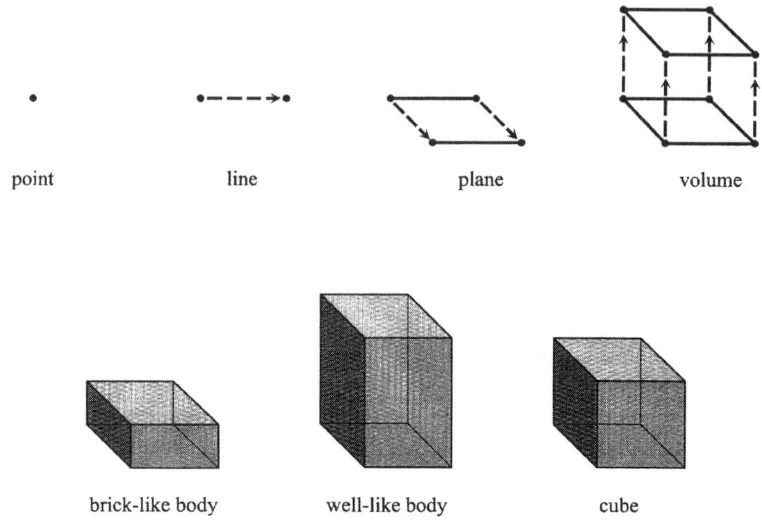

Fig. 4.6 The point as the generative principle of rectilinear bodies according to the Ikhwān.

be seen on its own, but only as it lies between two bodies, like the common surface between fat and water. An intelligible point, too, cannot be seen on its own, but only where a line is divided by imagination into two halves; wherever a division is indicated the point is marked there. And know, O brother, that if you imagine this point moving in one direction, an imaginary straight line would occur in your mind. And if you imagine this line moving in a direction other than that toward which the point has moved, an imaginary plane would occur in your mind. And if you imagine this plane moving in yet another direction to those of the point and the line, an imaginary body that has six square planes with right angles, that is, a cube, would occur in your imagination. If the distance traced by the movement of the plane is shorter than that traced by the line, a brick-like body (*jism labiniyy*) will occur; and if it is longer, a well-like body (*jism bi'riyy*) will occur; whereas if they are equal, a cube will occur.

And know, O brother, that every straight line conceived in the imagination must have two ends which are its two extremes; they are called the "two imaginary points." If you imagine that one of the two points has moved, while the other stayed still, until it returns back to the point where movement began, an imaginary circular plane will occur in your mind. The still point then becomes the center of the circle, while the moving point marks, in your mind, by its movement the circumference of the circle. Then know that the first plane that occurs is the quarter of the circle, then the third, then

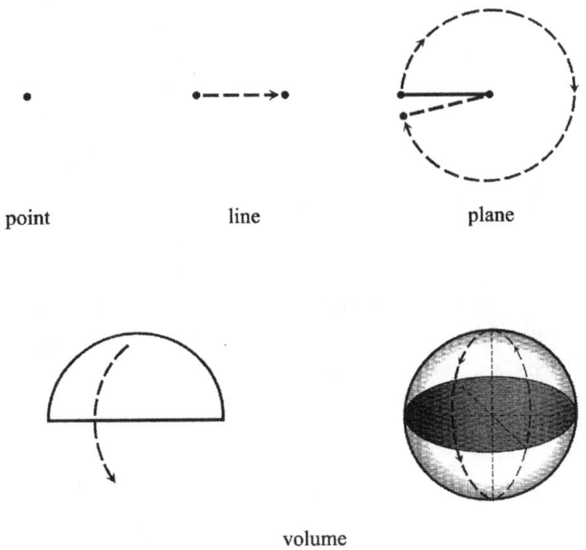

Fig. 4.7 The point as the generative principle of spherical bodies according to the Ikhwān.

one half, then the circle. If you then imagine that the curved line, which forms half of the circumference of the circle, is moving while its two ends are still, until it returns to where it began moving, a spherical body will occur in your mind. So it is clear to you, from what we have said, that intelligible geometry is the reflection on the three dimensions, which are the length, the width, and the depth as abstracted from natural bodies.[11]

And know that many of the *muhandisīn* (geometers, architects) and scientists imagine that these dimensions, I mean the length, width, and depth, have forms that exist by themselves, without knowing that this existence is either in the substance of the body or the substance of the soul. To the dimensions, these substances are like matter *(hayūlā)*, and in the substances the dimensions are like form *(ṣūra)*, detached by the thinking faculty from the sensible bodies. If only they knew that the ultimate aim of studying mathematical sciences is the training of the students' souls to be able to abstract, through the senses, the forms of the sensible bodies and to conceive their essences by the intellectual faculty, so that when the sensible bodies disappear from contact with the senses these forms—which have been transmitted from the senses to the imagination, and from the imagination to the intellectual faculty, and from the intellectual faculty to the memory—remain formed in the substance of the soul. The soul, then, when turning to itself, will dispense with using the senses in perceiving the information. It will find the forms of all information in its own substance.[12]

This mode of geometrical reasoning of the fundamental order in space is based on the necessity for a principle, following the divine model wherein the Essence projects itself into the world as the main cause of existence and principial element of order. And once the first principle is identified it is only logical to search for a progressional order of manifestation that explains the existence of the complex from the simple and the manifold from the one. This is not limited to the field of geometry, of course, but extends to other sciences, culminating with metaphysics whose principle is seen to be the Principle of all principles. The understanding of space and spatial manifestations as three dimensional projections from a principial point can be seen as a foundational concept that shaped premodern spatial sensibility, resulting in a consistent sense of ordering across geo-cultural, temporal, and typological variation.

Architecture and the Sacred

The 'sacred' is a key concept in modern discourses of the symbolism of premodern architecture. Cosmology and architecture meet on sacred grounds, and forms that embody cosmological ideas invoke the sacred. The sacred constitutes the complex religious context of spatial ordering.

The Sacred and the Profane

Our current understanding of the nature of the sacred is profoundly shaped by the works of two influential scholars, Rudolf Otto and Mircea Eliade.[13] Ever since the appearance of Otto's seminal work *Das Heilige* (The Holy) in 1917, the notion of the holy has come to denote primarily the mysterious, the numinous, the ineffable, and the nonrational.[14] Otto's main challenge, when he wrote the book, was explaining the ineffable nature of the *mysterium* to a dispassionate and intensely rational German audience. As a theologian and historian of religion, Otto was clearly writing from a defensive position. Rationalism, the prevailing intellectual condition of the time, relegated to the irrational anything that was inexplicable in rigorous and scientifically coherent terms. Throughout the text Otto was at pains to make an argument for the nonrational or suprarational as an alternative to the irrational. His main focus was the psychology of the holy; he wanted to make a case for "the *feeling* which remains where the *concept* fails."[15] Otto's struggle was clearly reflected in the English edition (first appeared in 1923), which bore the following clumsy title: *The Idea of the Holy: An Inquiry into the Non-Rational Factor in the Divine and Its Relation to the Rational.* Clearly, the perceived conflict between the rational and nonrational aspects of religion was the framework within which the notion of the holy was understood and defined. In this intellectual context, Otto presented the *mysterious* core of the holy as the "wholly other." He saw it as the inexplicable essence that lies "beyond the sphere of the usual, the intelligible, and the familiar,"[16] as the incomprehensible alien that fills the mind with "blank wonder and astonishment."[17] "The truly mysterious object is beyond our apprehension and comprehension," he wrote, "not only because our knowledge has certain irremovable limits, but because in it we come upon something inherently 'wholly other,' whose kind and character are incommensurable with our own, and before which we therefore recoil in a wonder that strikes us chill and numb."[18]

In *The Sacred and the Profane* Eliade developed Otto's perspective, while maintaining a polarized understanding of the sacred. "The first possible definition of the *sacred*," he wrote, "is that it is *the opposite of the profane*."[19] Eliade articulated his understanding of the sacred as a "hierophany," that is, "something sacred that shows itself to us."[20] It is the manifestation of divine *otherness* at the human level of existence. Eliade saw the whole history of religions to be constituted by the manifestations of a wide range of sacred realities. Unlike Otto's preoccupation with the psychology of the sacred, however, Eliade focused on its mythico-symbolic functions and spatio-temporal conditions. Central to his thesis was his conception of the heterogeneous nature of space and time. For *homo religiosus*, Eliade argues, space and time are not homogeneous extension and duration. The sacred marks "interruptions and breaks" in their continuum by disturbing the predictability of the natural processes and familiar conditions of existence. Where and when this happens, the sacred manifests itself in spatial or temporal forms that reconfigure the profane's nondifferentiated continuum. Thus the manifestation of the sacred disrupts the homogenous continuity of space and nullifies the linearity of history. With the manifestation of the sacred, space becomes centralized and time cyclical: the amorphous becomes structured according to archetypal paradigms.[21]

Eliade's theoretical analyses were of a wide appeal. Not only anthropologists, ethnologists, historians of religion, and social scientists were keen to explore Eliade's polarity, but also architects, architecture historians, and theoreticians were also quick to appropriate it into the field. Through the agency of function, the sacred-profane polarity was uncritically extended to architecture. A functional-typological split between religious and secular architecture was convenient and easy to define. It made sense to academics as well as professionals, leading to polarized historical and theoretical studies as well as design practices. While buildings, landscapes, and settlements, in general, are interpreted in sociopolitical, environmental, and aesthetic terms, only those that serve religious purposes are seen to accommodate the sacred and the symbolic. It is not the act of making or the fabric of the made object per se that legitimates the presence of the holy but rather the human function that takes place within.

The fabric is seen merely as a material support for the human activities. In such a polarized setting, a place of worship, for example, presupposes a mode of engagement with architecture that is different to, say, a marketplace. One seeks the holy; the other ignores it.

A Sacred without a Profane

Insightful and illuminating though it may be, Eliade's polarized approach to the study of the holy in general, and of sacred space, landscape, and architecture in particular, involves many problems. The most significant challenge the Islamic tradition poses to both Otto's and Eliade's polarized thinking is the lack of polarity. Neither was the sacred defined with reference to an antonym, nor was it viewed to pose a challenge to rationality. Translators of Eliade's works into Arabic face the difficulty of finding not only that critical definer, the elusive *profane*, but also the equivalents to many of his terms and concepts.[22] The "profane" has been rendered in various translations as *al-ʿādī* (the "ordinary") and *al-mudannas* (the "impure," the "desecrated"), yet neither forms a polarity with the sacred that is traceable in premodern Arabic literature. True, the absence of a term for the "profane" does not automatically rule out the availability of the concept or the possibility of a polarity, but it does make it difficult to maintain the sharp distinction Eliade draws between the two domains.

Apart from invoking the numinous, the notion of the "sacred" in the Judaeo-Christian tradition, as Otto explains, is generally understood as a moral attribute. It is the "completely Good." The "holy," Otto writes, stands for "the absolute moral attribute, denoting the consummation of moral goodness."[23] The Arabic term for the "sacred," *muqaddas*, while sharing the association with the numinous, denotes the idea of "purity." "Purity" was understood as a bodily rather than moral attribute, although one can make an argument for the overlap between spiritual purity and moral goodness. Purity, in premodern Islamic texts, signified proximity to the primordial nature *(fiṭra)*, the initial condition of one's being, hence its association with sacredness. The Muslim version of the narrative of Moses and the burning bush illustrates this subtle difference. The Quran recounts the events in a similar way to the Torah: "Verily I am your Lord. So take off your shoes, for you are in the sacred valley, Ṭuwa" (20:12). Eliade refers to this event in order to illustrate his fundamental concept of 'spatial heterogeneity.' He cites the story to show that there is "a sacred space, and hence a strong, significant space" that stands in contrast to other amorphous spaces "that are not sacred and so are without structure or consistency."[24] Taking off the shoes, Eliade argues, marks the discontinuity of the profane space that is ruptured by the manifestation of the sacred.

Following Eliade's logic one would expect to find in medieval theological texts that interpret this Quranic verse some treatment of the spatial qualities of the "sacred valley, Ṭuwa," or some preoccupation with the characteristics that distinguish it from the nonsacred surrounds, or at least a curiosity to delimit the sacred from the ordinary. But this is not the case. Instead, we find a debate of what Moses' shoes were made of that made them unfit for a holy presence. It was not spatiality that seemed to be the main concern, but rather the purity of the body. The sandals, some argued, were made of dead donkey's skin, and that contravened the purity of the sacred.[25] The contact of the bare feet with the ground was considered necessary for the direct contact with the numinous source and associated blessing conferred on that site. Of course, this does not devalue Eliade's insights and compelling interpretations but points to the necessity of identifying clearly the reading protocol that is being followed. While Eliade passionately argues that this is how premodern *homo religiosus* has always and everywhere conceived of and understood his spatial condition, his readings are in fact modern academic constructions that are often incongruent with the premodern perspectives they represent.

Medieval Arabic sources do not speak of "sacred" sites, landscapes, and cities as distinct from other types that are "profane," nor do they interpret spatiality in a dualistic frame of real and unreal, structured and amorphous, significant and insignificant. They also present us with scanty references to ritualistic practices of consecration that are associated with laying out buildings, settlements, or gardens. The absence of elaborate rituals in Islam presents yet another challenge to Eliade's interpretations. There are numerous references to astrological correspondences, but these are not presented as a part of consistent ritualistic practices for consecrating human acts of making. Apart from the reference to the "sacred valley" of Ṭuwa and the "sacred land" of the Jews, the Quran does not use the term *muqaddas* to identify sacred sites, not even the most *sacred* of all, Mecca. This does not mean, of course, that in premodern Islam there were no conceptions of the sacred or sacred sites but that the understanding and construction of the "sacred" itself was different. Conceptually, a "sacred" without a profane must necessarily be different from a "sacred" with a profane. Such difference presupposes a unique spatio-temporal understanding that constitutes the conditions of the sacred's modus operandi.

The Sacred as the Virtuous

In premodern Islam, we encounter an understanding of the scared as the "virtuous," meaning that which has special merits. The notion of the *virtuous* accounts for both the sacred and the profane in that all sites and places have virtues. The intensity and significance of the virtuous, however, vary from one place to another. The variations are hierarchically ordered and are charted through a unique form of conceptual mapping of holiness that is traced by Muslims on the territories they inhabit.[26]

The notion of the scared in relation to places, integral to premodern Islamic cosmology, was best expressed through the unique premodern genre of *faḍāʾil*. Literally meaning "merits," "virtues," or "excellences," the term *faḍāʾil* was used mainly as an adjective denoting the distinctive virtues of, or merits associated with, certain texts, individuals, places, or times. The intent is panegyric, while viewing things from a peculiar divine-human perspective for the purpose of laudation.[27] A large portion of the *faḍāʾil* texts is devoted to the distinctive virtues of provinces, cities, places, and monuments, which began to develop into a unique genre of literature from the twelfth century onward. In their early forms, dating from the first century of Islam, the *faḍāʾil* texts consisted mainly of a compilation of sayings attributed to the Prophet and his immediate companions. Later on, they developed into a unique style of historiography establishing a recognizable discourse.

Evidenced by its pervasiveness in most forms of literary expressions, there is no doubt that the concept of the '*faḍāʾil*' was central to the modes of thinking and seeing the world in premodern Islam. While the works of later Muslim scholars, such as Ibn ʿAsākir (d. 1175), al-Maqdisī (d. 1364), al-Suyūṭī (d. 1475), and al-Ḥanbalī (d. 1520), among many others, present the *faḍāʾil* in its most developed and sophisticated form of religious historiography, the works of chroniclers, geographers, theologians, mystics, travelers, and literary scholars throughout premodern Islam also incorporate, in a more or less conspicuous way, the perspective of the *faḍāʾil*. The ubiquity and legitimacy of the *faḍāʾil* discourse derives primarily from being as it were a by-product and a direct extension of the science of prophetic traditions (*ʿilm al-ḥadīth*).

The *faḍāʾil* is anchored in the Quran and the *ḥadīth*, which identifies and alludes to many sites of special significance. In the chapter of "The Fig," for example, the Quran says: "By the fig and the olive, by the Mount of Sinai, and by this land made safe" (95:1–3). These were interpreted as referring to four significant sites. The Damascene historian and *ḥadīth* scholar Ibn ʿAsākir explains that the "fig" refers to the mosque of Damascus; the "olive" to the mosque of Jerusalem *(bayt al-maqdis)*, "Mount Sinai" to the spot where God spoke to Moses, and the "land made safe" to Mecca."[28] Ibn ʿAsākir further elaborates these references by reporting a set of traditions that describe the eschatological significance of these sites. For example, a recurrent *ḥadīth* says that the construction of the mosque of Damascus was predicted long before it was actually built and that it will survive the destruction of the world by forty years.[29]

Jerusalem, al-Quds, Bayt al-Maqdis, and al-Bayt al-Muqaddas in Arabic, meaning "holiness," "the house of holiness," and "the holy house," is one of the most sacred sites in Islam on which numerous *faḍāʾil* texts were composed.[30] In these texts one encounters complex spatial conceptions presented through prophetic sayings and popular narratives. For instance, Jerusalem is said to be God's favorite spot on earth, toward which he glances twice a day. It is the center of the earth and the closest place to heaven by eighteen miles. Whether because of this spatial proximity or other cosmological reasons, paradise is said to "long" passionately for Jerusalem. In fact, Jerusalem itself is said to be an extension of the heavenly geography of paradise. Paradise is envisioned to be located directly above it, and those aspiring for a foretaste of paradise are directed to see Jerusalem. Along with Jerusalem, Mecca and Medina are consistently cited among the cities with heavenly connection (other urban centers are often selectively included). When the three most sacred centers, Mecca, Medina, and Jerusalem, are projected together a hierarchy of holiness is always evident. Visitors of Mecca are, for example, said to be forgiven and elevated eight steps, whereas visitors of Medina are to be forgiven and elevated six steps, while visitors of Jerusalem are to be forgiven and elevated four steps.

Through such as well as various other means the *faḍāʾil* discourse confers significance on places, monuments, and landscapes. It constructs the virtues of a particular site through a complex juxtaposition of various religious, cosmological, eschatological, and environmental references, while at the same time weaving together elements of vernacular history, sacred geography, religious rituals, and popular legends. Through such geo-mythical conceptions, the *faḍāʾil* enables, on the one hand, the construction of imaginative geographies that differentiates sharply between places, and on the other, the blurring of the boundary between the mythical and the real. It is through this blurring of spatiality that the *faḍāʾil* confers significance on places, buildings, and landscapes.

The Sacred and Difference

The concept of '*faḍāʾil*' reveals a discriminatory view of things based on God's own "preference" (*tafḍīl*, a derivative of *faḍāʾil*). The way things are materializes this divine preference. Things do not just happen serendipitously, but manifest in accordance with divine partiality, the logic of which hinges on the *necessity* of difference. In the overall scheme of creation, according not only to Islam but also to many other religions, different peoples, texts, and places are not of equal status. In the beginning was *difference*, and difference was never meant to be projected democratically. Difference was predicated on a preference, an absolute, nonnegotiable, divine preference. From this perspective, the *faḍāʾil* conception can be seen as an attempt to lay out spatially the matrix of differentiation and to reveal the pattern of divine partiality. It is as it were a literary act to inscribe the ontological foundation of difference.

'Difference' is, of course, a relational concept that requires a horizon of reference against which the other is to be differentiated. Naturally, the *faḍā'il* projects Islam as the horizon of reference against which divine preferences are identified and explained. The non-Muslim other occupies an awkwardly marginal position that is never in concordance with the order of things. The other is *de-placed.* The *faḍā'il* texts on Jerusalem, for example, relate an elaborate story of how the Christians' attempts to construct a monumental building over the sacred rock (where the Dome of the Rock was later built), long before the Islamic takeover, repeatedly failed. Three times their exquisite and highly adorned structure miraculously collapsed, forcing them in the end to consider a different site. It was not their architectural or engineering inadequacies that led to the repeated collapse but simply their religious otherness. The site, originally designated for Islam, could only tolerate an architecture that facilitates spatial practices that are in harmony with the Islamic creed of absolute unity. In this manner, the *faḍā'il* discourse founds not just difference per se, but a politicized difference. Difference is politicized through religious scenarios of encounters with the sacred, which take place in the blurred spatiality of the real and the imaginary, the earthly and the heavenly. It is a determinedly Islamic version of geo-politics wherein God, along with the Muslims, acts as a central figure in the plotting, unfolding, and staging of events.

Design and Cosmic Paradigms

Apart from the typological split of the sacred and the profane, Eliade's approach enabled a relocation of architecture into a context where the determinants of forms are not entirely human centered, where the divine is seen as an active partner in the act of siting and designing. Sacred places, Eliade argues, are not "chosen but rather discovered by religious man." "The sacred place in some way or another reveals itself to him."[31] In the absence of a direct revelation, holiness can still be invoked by human consecration through the enactment of certain religious rituals. Once consecrated, a sacred space becomes a defined, qualified, and significant space; it becomes an ordered space: a space with precise limits measured in accordance with cosmic paradigms. The act of ordering with reference to cosmic paradigms, Eliade argues, invokes God's blueprint of the world, the universal pattern of creation, and the principal elements of determination that emerged out of the primordial chaos.

We know very little about the procedures involved in designating spaces and the demarcation of their limits in the Islamic tradition. The designation of a sacred zone by setting out demarcating boundaries can be traces in some premodern Islamic references. Many early places for prayer were said to have

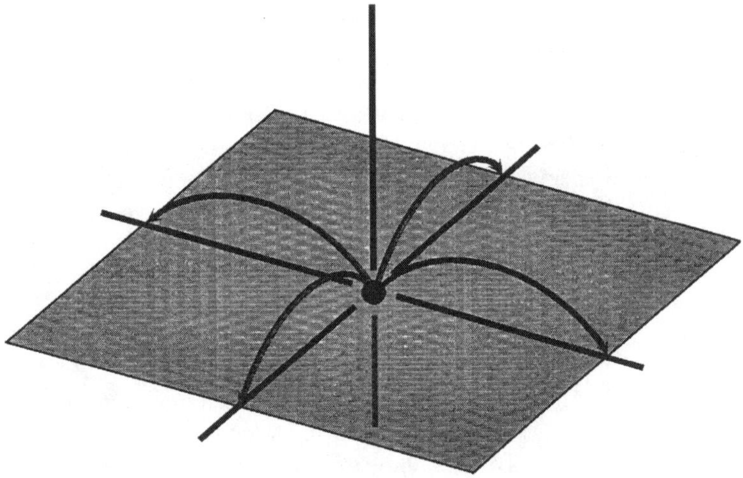

Fig. 4.8 The demarcation of the central space in the layout of the early cities of al-Kūfa and al-Baṣra according to early Islamic sources.

merely lines drawn in the sand: "[O]nce the worshipper had stepped over the boundary thus demarcated he was within a sanctified area in which all the Quranic taboos governing ritual purity were in operation."[32] The procedures involved in laying out the early cities of al-Baṣra and al-Kūfa are significant. Medieval sources say that the mosques of these two cities were first laid out, forming the centers of their internal open spaces *(al-ṣaḥn)*. The boundaries of these internal spaces were determined by shooting arrows toward the four directions from their central points (i.e., their mosques).[33] The space thus determined was then marked out by a ditch, and the residential zone was established beyond this boundary. Chroniclers also say that the city of al-Kūfa has four openings or gateways.[34] A similar planning pattern was reproduced in the city of Baghdād (762), the famous city of al-Manṣūr, the so-called City of Peace, which was laid out in concentric circles about a central point. The mosque and the caliph's residence occupied the center, which was surrounded by a circular residential zone that encompassed a large empty space. The surrounding residential belt was punctuated by four fortified gates, marking the ends of two perpendicular axes intersecting at the center of the city. The city was made circular, historian al-Baghdādī (d. 1071) says, so that all residents are equally related to the caliph who resides at the center.[35]

Premodern Islamic chronicles provide other references to significant sites. They tell us that many buildings, cities, and gardens were erected consciously on sites that were chosen with the help of scriptural, cosmological, or other supranatural references. Some settlement sites were chosen on particular auspicious dates determined astrologically or with reference to traditions that predict the location of the settlement or the destiny of the ruler. Through this religious reference the act of decision making is mediated by a divine reference giving the site or settlement special significance. For instance, the mount on which the mosque of Ibn Ṭūlūn was built in Cairo was believed to have been a blessed place, where a prayer was always heard and where Moses conversed with the Lord.[36] The foundations of the city of Baghdād were said to have been laid at a particular date chosen by the astrologer Nabūkhat, and so was the founding of Kāshān and Cairo.[37] The renowned Egyptian historian al-Maqrīzī (d. 1442) relates an amusing story concerning the founding of the city of Cairo. After the astrologers chose the appropriate dates to dig for the foundation of the city's protecting wall, all the workers were instructed to commence work simultaneously when the astrologers observed the ascent of the stellar reference. So that the workers would know when to start, bells were attached to a rope stretching throughout the work area. But in this instance things went wrong. A crow landed on the rope, ringing the bells and prompting the workers to begin under the ascendancy of the wrong star, which gave Cairo its current name, al-Qāhira. Regardless of the historical reality of such anecdotes, they nonetheless reveal a particular preoccupation with some kind of suprarational references to legitimate human preferences.[38]

Orientation

Acts of orientation reveal how people's spatial sensibility is influenced by the working of the sacred. The Quran describes Muḥammad's inner desire for a new sacred center *(qibla)*[39] to pray toward other than Jerusalem, in response to which the Kaʿba was chosen. The Quran says: "We have seen the turning of your face to heaven. We shall therefore make you turn toward a *qibla* that pleases you. So turn your face toward the Holy Mosque, and you (O Muslims), wheresoever you may be, turn your faces toward it" (2:144). Since this event all mosques have been oriented toward the Kaʿba, the divinely chosen center of the Islamic world. And since Muslims can pray practically anywhere, other buildings, such as tombs or schools, are often provided with niches for prayer and aligned with the direction of the *qibla*. Even entire cities, with more or less orthogonal street plans, are sometime laid out facing the direction of Mecca so that religious buildings could thus be aligned with the street patterns.[40]

It might seem a simple liturgical practice; however, in premodern Islam orientation took on cosmic significance. Orienting a built form toward the sacred center means positioning one's self and space on the grid of the divine map of holiness. Facing the Kaʿba can thus orient one's mind toward the celestial archetypes that lie directly above it. A marginal commentary on Ibn ʿArabī's treatise on Transcendent Unity *(Risālat al-Aḥadiyya)* articulates a fivefold structure of the *qibla*.[41] The first *qibla*, it says, is the niche *(al-miḥrāb)* of a mosque; the second is the Kaʿba; the third is the Frequented House *(al-bayt al-maʿmūr)*; the fourth is the Throne *(al-ʿarsh)*; and the fifth is the Footstool *(al-kursī)*. The niche is the *qibla* of the soul *(al-nafs)*; the Kaʿba is the *qibla* of intention *(al-niyya)*; the Frequented House is the *qibla* of understanding *(al-fahm)*; the Throne is the *qibla* of the heart *(al-qalb)*; and the Footstool is the *qibla* of the intellect *(al-ʿaql)*. Orienting oneself, and by extension a built form, toward the *qibla*, can thus be seen as establishing a horizontal link with the center of the world and a vertical link with the celestial centers marking the axis of the world. Orientation, in this sense, is an act of integration that establishes a way of return from the fragmented to the unified, from the complex to the simple, from the accidental to the essential, and from the many to the one.

In addition to the liturgical alignment with the direction of the *qibla*, there is a spatial alignment with the cardinal and intercardinal directions as marked by the sun's trajectory in its diurnal and annual journeys. Although the spatial ordering of Islamic buildings and landscape, as we have seen, emphasize the cross of directions with elements such as four doors, openings, *īwāns*, gateways, channels of water, or two major perpendicularly intersecting thoroughfares, the cross of directions is not always aligned with the cardinal points. The visitor to the bazaar of Iṣfahān and Masjid-i-Shāh, for example, would notice that in the buildings clustered alongside the spinal route of the bazaar there are numerous spaces of varying sizes and significance with clearly identified centers and cross of directions. Given the crookedness of the route, to which many of these spaces are parallel, and the orientation of the city, it is evident that the quadrature marked by their cruciform pattern has little to do with the cardinal directions. A cruciform planning pattern reappears in the Masjid-i-Shāh with which the bazaar ends. The orientation of the mosque masterfully ruptures the north-south axis of the city in order to face Mecca, yet the center and the cross of directions are clearly marked not only in its central courtyard but also in the other minor spaces that form parts of the overall composition.

The alignment of buildings with the cardinal points is, as many studies have shown, a well-established ancient practice. In the Hindu tradition, for instance, there are building manuals that prescribe the rituals of laying out buildings in accordance with cosmic geometry. Traversing the parameter of the celestial space, the sun determines the four extremities of spatial extension—east, west, north, and south—and the four nodes of the temporal cycle—the four seasons and the four temporal measures, year, month, week, and day. These spatial and temporal determinations are "married in the motions of the solar orb."[42] Marking the center and the cross of directions through architecture is shown to have been understood as tracing the order of celestial geometry. By way of correspondence with the solar cycles, the plan of a building becomes, so to speak, an architectural crystallization of temporal cycles, a cosmic graph, a projection of the celestial geometry, a geometrization of time, and a coagulation of time in spatial form.[43]

The Islamic tradition left no textual references to consistent rituals of laying out buildings, settlements, and landscapes similar to those found in other traditions. Yet Muslims seem to have adhered to the same sense ordering and spatial sensibility, appropriating the ancient practices into their religious framework. With the orientation toward Mecca taking priority, the cruciform pattern acquires a new significance, one that is anchored in human spatiality viewed, as we have seen, as the original idea and the divine paradigm of cosmic structure. This can be traced in the temporal notion of '*waqt*.'

Al-Waqt *and Spatial Order*

Islamic prayers occur at certain times (*awqāt*, singular, *waqt*) of the day, the accuracy of which is essential for the prayer's validity. Defining those times is a spatial as much as it is a temporal exercise. *Al-waqt*, literally "a period/point in time," denotes, according to Ibn ʿArabī, a designation *(taqdīr)* in something that in itself does not admit what is being designated. It is an assumption, in other words, as is the case when one assumes a beginning, middle, or end in a sphere, while the spatiality of the sphere does not admit any of these definitions. With reference to a prophetic tradition that describes time *(zamān)* as being circular in form, Ibn ʿArabī argues that *al-awqāt*, as temporal assumptions, are meaningful only with reference to both human spatiality and man's centrality in the world. It is the correlation of the stellar movements with human spatiality that establishes the spatio-temporal order of the world. Ibn ʿArabī explains:

> When God created the *aṭlas* sphere and it revolved, the day was not yet determined, nor did it have a designated form. It was as the water of a jug when the water was still in the river before being in the jug. But when God designated in the *aṭlas* sphere the twelve divisions, which were precisely timed, and called them "signs" *(burūj)* . . . , set an individual standing [in the center] about whom this sphere revolved, and rendered this individual with sight whereby he observed those designated divisions by means of the signs rendered in them, these supposed divisions were distinguished from one another by those signs which are made as allusions to them. This individual sighted one of those designated divisions, I mean, the sign, then the sphere rotated with that sign, on which the spectator had already fixed his sight. The sign

vanished from his sight, but he never ceased standing in his position there until that sign had returned to him. It was only then that he knew that the sphere has revolved one circuit, in relation to this spectator and not to the sphere, so we called that circuit "*yawm*" (day). Then afterward God created an immense luminous planet in the fourth of the seven heavens. He called it in the Arabic tongue "*shams*" (sun). In the sight of the spectator, the *aṭlas* sphere rose with the sun from behind the veil of the earth, upon which this spectator is standing, so he called that place of the sunrise "*mashriq*" (east),[44] and the rising "*shurūq*" (illumination); because that luminous planet rose from it and illumined the atmosphere wherein the spectator was standing. His sight never ceased following the movement of that planet [as it rose] until it coincided with his position; he called this coincidence "*istiwāʾ*" (resting). Then the planet began descending from its resting position towards the right-hand side of the spectator,[45] with regard to the spectator and not the planet itself, as we have already said. He called the beginning of its dissociation from its resting position, in the eye of the spectator, "*zawāl*" (vanishing, disappearing) and "*dulūk*" (moving from the center of heaven). The spectator kept on following it by his sight until the body of the planet disappeared; he called its disappearance "*ghurūb*" (setting), and the place where his sight saw the sun disappearing "*maghrib*" (west).[46]

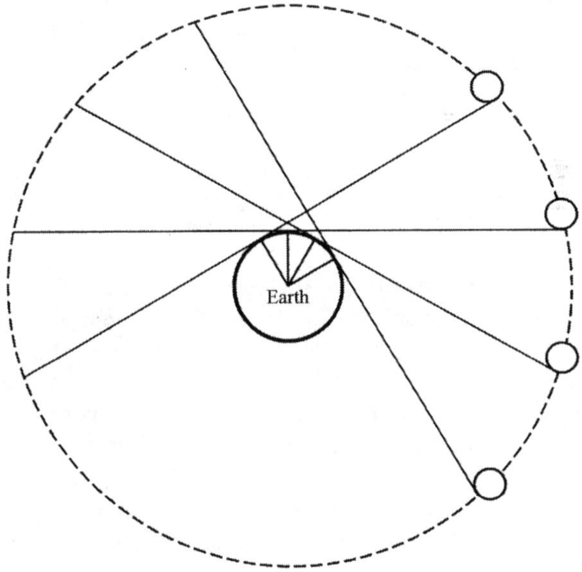

Fig. 4.9 The moving horizon in relation to the sun and the viewer's position.

In its continuous movement around the stationary earth, the sun, according to this depiction, neither rises nor sets. Its perpetual movement reflects the non-differentiated motion of the *aṭlas* sphere, which, according to al-Qāshānī, is but a temporal expression of the undetermined duration of the subsistence of the divine Essence from "preeternity of preeternities" *(azaliyyat al-āzāl)* to "posteternity of posteternities" *(abadiyyat al-ābād)*.[47] The cardinal points are spatial determinations of the temporal differentiation of the flux of time into recurring cycles of prescribed diurnal or annual durations by means of a reference point. Man's centrality marks the reference point, without which there can be neither east nor west, nor can the cross of the directions marked by the solstitial and equinoctial nodal points be meaningful.[48] This is why Ibn ʿArabī considers 'direction' to be a relational concept, defined as neither existent nor nonexistent, neither spatial nor nonspatial, neither self-supported nor supported by other substances. 'East,' 'west,' 'north,' and 'south' are thus concepts identifiable only with reference to a given visual horizon established by a fixed point on earth. Man represents the fixed point, and his vision is what establishes the horizon. The determination of the cardinal directions coincides with the determination of *al-awqāt*. The directions qualify the nondifferentiated expanse of space, in the same way that *al-awqāt* qualify the nondifferentiated duration of time.

Long before Islam the Arabs also used human spatiality as a reference for qualifying the directions of space. This is evident in the Arabic names of the four winds that were astronomically determined, and with which the four sides of the Kaʿba were aligned.[49] *Al-qabūl*, the eastern wind, derives from *qubl*, "man's front"; *al-dabūr*, the western wind, derives from *dubr*, "man's back"; *al-shamāl*, the northern wind, derives from *shimāl*, "man's left-hand side"; and *al-janūb*, the southern wind, derives from *janb*, literally "man's side," referring indirectly to man's right-hand side.[50] The alignments of the Kaʿba with the four winds shows the primacy of human spatiality in identifying directions in space. This can also be traced in the shooting of the arrows in the four directions to determine the layout of the cities of al-Baṣra and al-Kūfa. The early chronicler and Quran scholar al-Ṭabarī (d. 923) defines these directions as the right-hand side, the left-hand side, the front, and the back of the archer, while the ninth-century historian al-Balādhurī relates them to the directions of the winds.[51]

Furthermore, from the very beginning Islam did not ascribe a special religious value to the sun. In fact, the sunrise, the zenith, and the sunset position are considered among the prohibited *awqāt* for prayer.[52] This is to avoid any coincidence between praying and these three nodal positions of the sun so that he whose *qibla* is aligned to the east or the west, or who happens to be standing directly below the sun, would not appear as if he were praying toward it.[53] This sensitivity is expressed in the verse: "Adore neither the sun nor the moon; but adore God who created them, if it is in truth him whom you worship" (41:37).

Whether it relates to the four cardinal points, to the four winds, or to the four directions of man, the tectonic quadrature can be seen to correspond with the quaternary order of manifestation and divine pattern of proliferation. It expresses a desire to impose a transcendental order on the human settings. The deployment of space from a central point along the axes of the cross of directions can be seen as an embodiment of the manifestation of the four creative attributes—Life, Knowledge, Will, and Power—from the unmanifest Essence, and a reenactment of the spreading out of the square of the earth from its unextended substance, the *turba*. The spatial and directional ordering marked by two horizontal axes can be seen to correspond to all nondirectional divine quadratures, the four supranatural principles—Intellect, Soul, Nature, and Matter—the four principles of Nature—heat, cold, dryness, and moistness—the four ideal elements *(arkān)*—fire, air, water, and earth—the four bearers of the Throne, the four supporters of the Footstool, and the four rivers of paradise. It can be taken to correspond to the four nodal points of the sun, the four intervals of the moon, the four seasons, the four measures of time—year, month, week, and day—the four kinds of signs—Igneous, Aerial, Aqueous, and Terrestrial—the four kingdoms—mineral, plant, animal, and human—to man's four humours—yellow bile, black bile, blood, and phlegm—and to his four natural forces—attractive, fixative, digestive, and repulsive. The marking of the four directions in space may also allude to the four spiritual masters *(awtād)* who guard these directions and to the four qualitatively different sets of Arabic letters. This set of correspondences is only a sample of a much wider range that can be found in numerous premodern Islamic sources.[54]

Heavenly Landscape

From his prison cell in the fortress of Fardajān, the great Muslim philosopher Ibn Sīnā (Avicenna, d. 1037) projected his burning desire to break out of the captivities of this world. In a visionary narrative, *Ḥayy bin Yaqẓān*, Ibn Sīnā revealed his longing to journey away from the wretched darkness of the Occident, and to set out for the enlightening beauty of the Orient. His spiritual guide, Ḥayy, a beautiful and youthful shaykh who shone with divine glory, came from al-Bayt al-Muqaddas, "the Most Holy Dwelling," with which the city of Jerusalem was identified. Ḥayy's job was ceaseless journeying, forever traveling around the universe to know its conditions.[55] In response to Ibn Sīnā's eager questioning, the shaykh provided fascinating descriptions of the cosmic terrains he visited, the various climes and their inhabitants, the Occident and its darkness, the spring of life, the cities, the mountains, the seas, and the perilous obstacles one has to overcome in order to reach the Orient. Eager to follow in Ḥayy's footsteps, Ibn Sīnā asked the shaykh to show him the way, to which the shaykh's surprising response was: "[T]he road is closed to you all."[56]

This was not always the case, of course. Ibn Sīnā was not the only one to speak of a spiritual journey or to visualize cosmic and paradisaical landscapes. His contemporary al-Maʿarrī (d. 1058) and later al-Suhrawardī (d. 1191) wrote equally fascinating accounts of the cosmic topography and environments. Ibn ʿArabī, too, needed no one to show him the way as he set out to explore the wondrous cities, landscapes, and inhabitants of the celestial world. His detailed and vivid descriptions reveal not only the richness and profundity of his spiritual experiences but also an intriguing sense of familiarity with the otherworldly things.

Such visualizations and experiences show that al-Ghazālī's gazing at the sky for psychic comfort and delight and for contemplating the wonders and beauty of God's design are only one aspect of the Muslims' fascination with heaven. Another more profound aspect is the imagining, constructing, and indeed experiencing the spatial order and architecture of the celestial world. This draws attention to the premodern Islamic understanding of spatial reality that includes both physical and spiritual spaces. In premodern Islamic sources one encounters a complex picture of the world where earthly geography occupies only a limited space. There are other vast, imaginal, yet real, terrains that even mainstream religious literature meticulously describes.[57] The Sufis have contributed significantly to the construction and propagation of such conceptions, while establishing a sense of actuality based on *real* experiences. Although it is difficult to make sense of such experiences in modern scientific terms, we cannot simply ignore them or reduce them to the "spiritual," "mystical" or "mythical" categories, in order to sharply distinguishing them from physical reality. For the experiences they unveil have spatial and bodily dimensions connected with the architecture and landscape of these ethereal terrains as well as socio-religious values.

"Where has the protected rider come from?" Ibn ʿArabī was asked upon arriving at the gateway of the celestial world. "From the land of the occidental body," he replied.[58] These experiences seem to be of an imaginative nature, combining spirituality and materiality in creative ways. Although spatiotemporal, they do not conform to the commonly known laws of nature. They have their own "natural" laws and corresponding spatio-temporal conditions. To what extent the visualization of such terrains had influenced the conceptualization and making of earthly spaces in architecture and landscape is a question that is yet to be adequately explored. The paucity of such studies in this area limits our understanding of this potentially rich possibility.

The Land of Reality and the Cities of Light

When God created Adam, the father of mankind, from fermented clay, Ibn ʿArabī writes, a very small remnant of the clay was left over. From this remnant God created the palm tree *(al-nakhla)*, Adam's "sister," yet, a tiny speck, the size of a sesame seed, still remained. Within this speck God spread a vast land, so vast, indeed, that it included all creations, even the divine Throne and what it contains—the Footstool, the skies and the earths, all that is below the earth and all levels of heaven and hell. The proportion of all of these to the vast land is as a ring thrown in a limitless desert. This is the Land of Reality *(arḍ al-ḥaqīqa)*.[59] Access to this land has certain protocols; Ibn ʿArabī explains:

> On that land there exist forms *(ṣuwar)* of wondrous formation and beautiful stature; they stand at the entrances to the avenues that overlook this world in which we are, its earth and heaven, its paradise and hell. When one of us wishes to enter this land, that is, the knowers from any kind they might be, human, jinn or angel or from the people of paradise, as long as he possesses adequate knowledge and has stepped outside his temple of flesh, he finds these forms at the entrances of the avenues, standing ready to carry out the responsibilities that God has charged them with. One of them runs to the new arrival; it clothes him in a dress appropriate to his rank, takes him by the hand, and walks with him through this land. The visitor goes wherever he wishes and reflects on God's artefacts . . . When the visitor wishes to return, his companion goes with him to the place where he entered; he salutes him, removes the dress in which he had clothed him and departs from him.[60]

The protocol of access and departure and the exchanges that take place at the threshold are intriguing. The process involves both continuity and transformation so that bodily engagements are not completely dispensed with. Ibn ʿArabī explains the mechanics of formal exchanges by way of an analogy. If one looks at a light source and squints one's eyes, he says, one will see a multitude of light rays connecting one's eyes with the luminous source. By gradually opening the eyes the rays contract back to the light source. The light source, he adds, represents the specific place of this land whence forms emanate; the pupil, the instrument of vision, represents the world; and the rays represent the new forms one's soul take on once detached from its body (during sleep, for example). One's intention to see the rays by squinting represents one's preparedness *(al-istiʿdād)*; the projection of rays represents the emanation of forms upon such preparedness; and the contraction of the rays represents the return of forms to the land upon the cessation of one's preparedness.[61]

The Land of Reality has been interpreted by modern scholars as an allegorical reference to the "imaginal world," the world where "all the essential realities of being . . . are manifested in real images."[62] It is an intermediary world between the physical and the metaphysical, the plane where spirits and bodies exchange qualities: spirits become materialized and bodies spiritualized. As described by Ibn ʿArabī, it is the land of great wonders, where many rational impossibilities and absurdities exist, where reality reveals itself in many fascinating forms and whence all imaginable forms emanate.[63]

The Land of Reality is the *qibla* of the Sufis, the place in which their active imagination is anchored. Those who visited this land reported what they had observed and learned there. They say that unlike things in our world, all things on that land are alive and endowed with a rational faculty. One can converse with, and learn from, gardens, animals, and minerals. "He passes near no stone, no tree, no village, nothing whatsoever," Ibn ʿArabī reports after a visitor, "without talking to it, if he wishes, as a man speaks with his companion. They have different languages, but this land has the characteristic of giving to all who enter it the understanding of all the languages that are spoken on it."[64] And while everything in our world is ephemeral, mortal, and mutable, things there are permanent, immortal, and immutable. This land contains many worlds, among them one made exactly in the form of our phenomenal world. If a Sufi observes this world he would see himself in it, as confirmed by the Prophet's companion Ibn ʿAbbās.[65]

Within this land there are many places with distinct characteristics. There is a place where everything—from minerals to fruits, to men, and so on—is made of red gold. The shapes and forms of trees and fruits are exactly the same as ours, but they are golden; their taste and fragrance, however, are far superior. Fruits are ornamented and decorated so beautifully that they can hardly be imagined. It is one of the many wonders of this land that if one picks a fruit from a tree another fruit in a ripe state instantly replaces it. Other spots are made of silver, of white camphor, and of saffron. The most exquisite place, however, is that of saffron "compared to whose women the Houries of Paradise fade into insignificance."[66] Time differs from one spot to another—a moment in one may equal a year in another.[67]

As for the architecture of this land, the inhabitants there can either build using the same tools and methods familiar to us or else build by mere imagination and intention. They have cities called the "Cities of Light" *(madāʾin al-nūr)*, whose structure is wondrous. They also have a Kaʿba, which they circumambulate in the same manner as the people of earth. Although it has the same quadrangular form as the earthly Kaʿba, the one in this land is larger and unclothed. And unlike the earthly one, this Kaʿba salutes, and speaks to, its ambulants, benefiting them with sciences they do not possess. Their cities, the Cities of Light, are multistoried, thirteen in number, built one above the other. They have gates with vaults that are made of enormous stones of hyacinth. When these cities were built a particular spot was first chosen and then a small city with great fences was constructed. Afterward, towers that are higher than those of the small city were erected away from the sides of the first one. Extending the building in between these towers the first city was covered up. The new structure became like a roof to the first one. This roof was used as a ground, and another city greater than the first one was built upon it. Building layer upon layer in this manner, thirteen stories were constructed.[68]

Ibn ʿArabī's diagrammatic representation of heaven and earth, already discussed, bears striking similarities to the Cities of Light, which can also be seen in premodern iconographic depictions of the celestial, terrestrial, and infraterrestrial worlds.[69] Seen as an extension to this land, the earthly environment provides the necessary reference for the transformed spatial experiences, which can be traced in the popular literature concerned with the modes of living in the hereafter.

The architect of the cities holds a privileged status in this land. Ibn ʿArabī reports after one visitor, who met many of the land's kings, that once he met a distinguished person sitting next to one of the kings. The person's gestures and movements attracted the attention of the visitor, so he asked the king about his status in the kingdom. The king smiled and asked: "Did you like him?" "Yes," the visitor replied. The king said, "He is the *miʿmār* (architect) who builds for us the houses and the cities, and everything you see is the traces of his work."[70]

The Kaʿba: The First House

According to premodern Islamic sources, Mecca was the omphalos of the earth, and the Kaʿba was God's first house of worship. Being, so to speak, the first divine-sponsored architectural project, the Kaʿba is a key element in the interplay of cosmology and architecture. With the help of angels, Adam is said to have been the builder of the primordial house *(al-bayt al-ʿatīq)*. During the deluge, however, the Kaʿba was raised to heaven, and it was Abraham who later rediscovered the site and reerected the sacred house under God's order and guidance. The Quran says: "And when we prepared for Abraham the place of the house" (22:26).

Traditional narratives elaborate on this, explaining that after the disappearance of the Kaʿba during the deluge, God ordered Abraham to reestablish the house of God. Not knowing where he should do so, Abraham asked: "O Lord, but where?" God replied: "We shall show you." God sent him a speaking "cloud" *(al-sakīna)* that directed him to the sacred spot.[71] Taking on the form of the house, the *sakīna* said: "O Abraham, your Lord orders you to design according to the measure of this cloud"; in another source: "O Abraham, take from the land according to my measure, with no increase or decrease";[72] Ibn ʿArabī reports: "Build according to the measure of my shadow."[73] Abraham traced precisely the form of the cloud, and with the assistance of his son Ishmael rebuilt the house as he was instructed. The Quran confirms: "And when Abraham and Ishmael were raising the foundations of the house (Abraham prayed): our Lord, accept this from us, you are the Hearer, the Knower" (2:127).

Al-sakīna, the divine agent that selected the sacred site and delivered the heavenly model of the Kaʿba, denotes the ideas of "centrality" and "peace." The name derives from *sukūn*, literally "stillness," and has been used in the Quran to denote the ideas of "repose," "peacefulness," and "certainty": "He it is who sent down *peace of reassurance (al-sakīna)* into the heart of the believers" (48:4).[74] It also relates to *sakana*, to "dwell," a meaning that alludes to the heart as God's "dwelling," the center of repose, the spring of certainty. The quest for principle is implicit in this term, since stillness is the quality par excellence of the center, the motionless mover, the unvocalized cause of utterances *(sukūn)*, and the immutable principle of change. As a visible embodiment of the *sakīna*, the Kaʿba becomes the heart of the world, the house of stillness, the locus of great peace, and the immanence of divinity at the center of the world.[75]

In Sufi terms, the Kaʿba's cube-like form is a crystallization of the cube of man. It is an embodiment of the human as well as cosmic spatial structure and a visible manifestation of the three-dimensional cross. Its four *arkān* correspond to the human nature, its six faces to the human figure, and its three dimensions of length, breadth, and depth to the human body. The form of the Kaʿba is also seen to correspond to the twenty-eight mansions of the moon and,

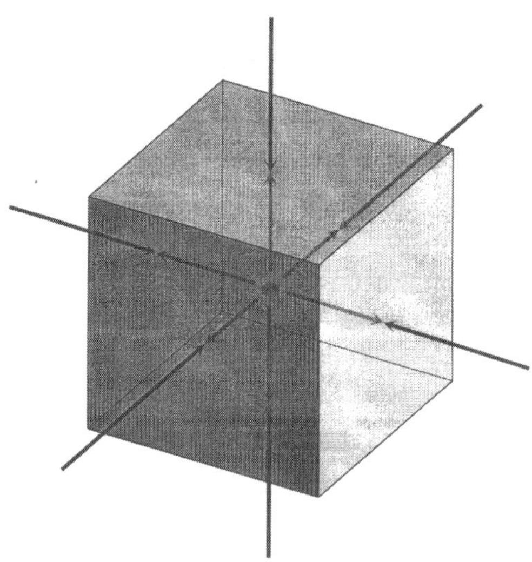

Fig. 4.10 The cube of the Kaʿba crystallizing the spatiality of the human presence.

consequently, to the twenty-eight letters of the Arabic alphabet. Ibn ʿArabī says that the height of the Kaʿba is twenty-eight cubits, twenty-seven cubits to the roof level, and one cubit for the parapet. Every cubit corresponds to a designation of divine order *(amr ilāhī)*. These designations, he says, are analogous "to the stations of the heart, traversed by the planets of faith in order to manifest events that occur within the soul, and this corresponds to the mansions of the moon, [traversed] by the mobile planets in order to produce events that occur in the natural world."[76]

Premodern literature on the Kaʿba provides ample references to the notion of centrality, axiality, triplicity, and quadrature; to its agency in the spatial deployment and temporal differentiation; and to its significance in materializing the creative relationship between triplicity and quadrature. The rich and complex mythology of the Kaʿba shows how a built form can become an integral part of divine geography and a central element in a cosmic landscape.

The Center of the World

Built according to a divine model and on a divinely chosen site, the Kaʿba was viewed to mark the center of the earth. Renowned early scholar al-Kisāʾī (d. c. 805) affirms: "Know that the center of the earth, according to a tradition on the authority of the Prophet, is the Kaʿba; it has the significance of the navel of the earth, because of its rising above the level of the earth."[77] Described in the Quran as "the mother of towns" (*umm al-qurā*, 6:92), Muslims were reassured that Mecca was indeed the navel of the earth. Al-Azraqī (d. after 858) in his famous chronicle *Akhbār Makka* reports many traditions concerned with the origin and significance of the Kaʿba that had continued to be reported in various forms until the dawn of the nineteenth century.[78] Mecca-centered maps prevalent in premodern times depict the Islamic world as a gigantic wheel with Mecca as its hub. The regions of this world were often identified by niches oriented toward the center. The lines radiating to all the mosques on earth, which represent the spokes of the wheel, form the axes of orientation that converge on the Kaʿba.[79] This imagery can be traced in early literature. Describing the Kaʿba's geo-cosmic position the Ikhwān write:

> The house *(al-bayt)* in the middle of the holy mosque *(al-masjid al-ḥarām)*, the holy mosque in the middle of the sanctuary *(al-ḥaram)*, the sanctuary in the middle of al-Ḥijāz, al-Ḥijāz in the middle of the Islamic countries, is in the likeness of the earth in the middle of the atmosphere, the atmosphere in the middle of the lunar sphere, the lunar sphere in the middle of the [celestial] spheres. And those who pray in the horizons oriented toward the house are in the likeness of the planets in the spheres—their radiations are directed towards

> the center of the earth. And the rotation of the heavens with their planets around the earth is in the likeness of the rotation of the ambulants around the house.[80]

Marking the center of the earth, the Kaʿba assumes a significant cosmological function. It is the initial element from which the earth was "spread out" *(duḥiyat)*. A reported tradition says that the site of the Kaʿba was created two thousand years before the earth, which was spread out from it.[81] Kaʿb al-Aḥbār, the Prophet's companion, is reported to have said: "Forty years before God created the heavens and the earth, the Kaʿba was a scum on the water, and from it earth was spread out."[82] Al-Azraqī reports a tradition that says that the Kaʿba stands on the exact spot where Adam built the first temple, the foundations of which were laid by the angels deep in the seventh earth. It is said that Adam was ordered to build a house to glorify God in the same manner of the angels. So he was led to the spot where the archangel Gabriel struck the earth with his wing, revealing a "firm foundation" *(uss thābit)* in the nethermost earth. The angels filled up this pit with immense rocks until it became level with the surface of the earth.[83] Adam then laid out the first temple on the angelic foundation. Al-Yaʿqūbī (d. after 905) adds that the Kaʿba was once burned down, and in the process of rebuilding, the Qurayshīs dug deep to Abraham's foundations, of which they accidentally extracted a stone. But the stone immediately jumped back to its position, and a large serpent emerged and prevented the workers from reaching the building. Such narratives show the cosmological significance of the Kaʿba seen as the navel of the earth, the sacrum *(ʿajb al-dhanab)* of the body of the world.[84]

In mystical exegesis, the Kaʿba assumes another level of significance whereby it becomes a visible trace of the process of universal manifestation. In an interesting interpretation of the Quranic verse, in which Abraham addresses God, saying: "Our Lord, I have settled some of my posterity in an uncultivated valley near your holy house" (14:37), the celebrated Bosnian Sufi al-Bīrāmī (d. 1644), known as *shāriḥ al-Fuṣūṣ*, reads an architectural embodiment of the divine process of determination. The "uncultivated valley" *(wādī lā zarʿa fīhī)*, he says, refers to the state of nondetermination *(al-lātaʿayyun)*: the "valley" being the divine Essence, and "cultivation" being the manifestation of the names and attributes. Thus understood, the "uncultivated valley," on which the Kaʿba was built, becomes a reference to the state of Transcendent Unity *(al-aḥadiyya)*, the state of *virginity* not yet *cultivated* with the considerations of the names and attributes, nor with their relations and additions, nor with their effects and determinations.[85] As "the first sanctuary appointed for mankind" (3:96), the Kaʿba then becomes as it were an architectural trace of the first act of determination. It becomes a tectonic expression of the manifestation of the divine presence from the unmanifest principle of Being. The laying out of the square plan of the Kaʿba in the uncultivated, virgin valley corresponds to the differentiation of the creative divine quadrature—Life, Knowledge, Will, and Power—from the undifferentiated unity, and to the differentiation of the quadrangular form of the Throne from the undifferentiated primordial Light. With reference to the Quranic verse: "So let them worship the Lord of this house" (106:3), the Kaʿba has also been associated with the presence of lordship *(al-rubūbiyya)*. Its "planting" in the plantless valley, al-Bīrāmī writes, is a visible trace of "the determining of the state of lordship in the uncultivated valley of essential unity *(al-waḥda al-dhātiyya)*."[86]

The Cosmic Axis

At the center of the earth the Kaʿba reveals a vertical relationship with the center of heaven. Premodern sources speak of the correspondence the Kaʿba has with the Polestar, viewed as the center of heaven. Al-Kisāʾi says that traditionally it is believed that "the Polestar proves the Kaʿba to be the highest situated territory on earth, for it (viz. the Kaʿba) is opposite the center of heaven."[87] He further explains:

> In the center of this moving part of heaven (viz. the Great Bear) is a fixed star which does not move, and this is the Polestar, around which the Bear and the rest of the stars turn. People are agreed on this point that he who places himself opposite the Polestar has at the same time the direction of the Kibla, because this star is above the Kaʿba, without ever moving. The Bear may move somewhat, but the Polestar never does. If now the Polestar, which is the center of heaven, around which the other stars turn, is above the Kaʿba, this fact proves that what corresponds with the center of heaven is most likely to be the center of the earth; consequently the Kaʿba is the center of the earth.[88]

The correspondence between the terrestrial and celestial centers suggests a perpendicular cosmic axis that penetrates the terrestrial and celestial terrains, tying the Kaʿba to their respective centers. This perpendicular connection makes the Kaʿba the nearest point to heaven. "In no place," says the Prophet's wife ʿĀʾisha, "I ever saw heaven nearer to earth than I saw it in Mecca."[89] This is also expressed by the tradition that says that Mecca "is situated twelve *mīl* nearer to heaven."[90] The cosmic axis that passes through the Kaʿba connects it to its infraterrestrial, celestial, and supracelestial counterparts. A prophetic tradition describes the location of the Kaʿba as being at the midpoint of an axis that penetrates the seven heavens and seven earths, marking at each level a central point whereupon stands a house similar in structure and sacredness to that of the Kaʿba. Al-Azraqī writes: "The Apostle of God said: this house is one of

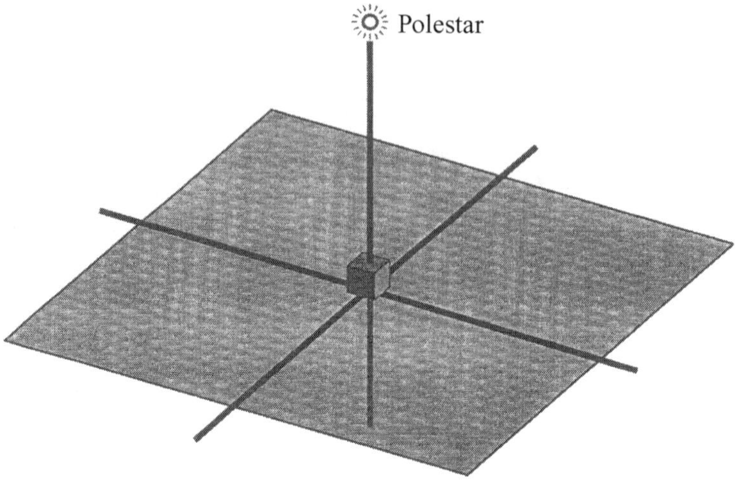

Fig. 4.11 The Kaʿba's axial relationship to the Polestar according to al-Kisāʾī.

fifteen, seven in the heavens up to the Throne and seven down to the limits of the lowest earth. The highest situated one, which is near the Throne, is the Frequented House *(al-bayt al-maʿmūr)*. Every one of these houses has a sacred territory, like that of the Kaʿba. If any one of them fell down, the rest would fall down, one upon the other, to the limits of the lowest earth. And every house has its heavenly or earthly worshippers, like the Kaʿba."[91]

The cosmic axis is the channel through which the higher cosmic entities pass their qualities onto their lower replicas. The Kaʿba is a visible replica of the Frequented House, also referred to as "*al-ḍurāḥ*," the Kaʿba's highest celestial counterpart, which in turn is a replica of the supracelestial model, the divine Throne.[92] A popular tradition says that "the house which is in heaven is called "*al-ḍurāḥ*"; its form is similar to this sacred house; if it falls, it would fall upon the house."[93] It is said that the Prophet saw it during his miraculous ascension and afterward described it to his companions as being located directly above the Kaʿba and directly below the Throne, and as being as sacred to the inhabitants of heaven as the Kaʿba is to the inhabitants of earth.

Ibn ʿArabī explains the unchanging axial relationship between the Kaʿba and its celestial counterpart, *al-ḍurāḥ*. He says that *al-ḍurāḥ* is located in the seventh heaven, which is standing still, as are the rest of the heavens. God made these heavens firm and settled; they are to us as the roof is to a house, and that is why heaven is called the "uplifted roof" *(al-saqf al-marfūʿ)*. As for the spheres *(aflāk)* in which the planets revolve, they are confined within these fixed heavens and not the heavens themselves. That is why *al-ḍurāḥ* does not move from its axial position opposite the Kaʿba.[94]

Axiality is associated with the idea of the mountain, seen as the natural place of communication between heaven and earth.[95] In a sense, the mountain, a primordial symbol of centrality and loftiness, represents the cosmic pillar that stands at the center of the earth and around which everything revolves. According to Islamic cosmogony, the position of the Kaʿba as the navel of the earth is evidenced by the fact that Abū Qubays, a mountain in the vicinity of Mecca, was the first mountain positioned on earth, and this is why Mecca is called the "mother of towns." Reporting after the Prophet, Ibn ʿAbbās says: "When, before the creation of heaven and earth, the Throne was upon the water, God most high sent a soft wind that struck the water, unveiling at the position of this house a piece of rock like a dome. God stretched out the earths from underneath it; it swayed and swayed again, so God most high pegged it by mountains, and the first mountain placed therein was Abū Qubays, and that is why Mecca was called the 'mother of towns.'"[96]

Spatial Deployment, Temporal Differentiation

Describing the form of the house that Adam first constructed on the Kaʿba's site, al-Azraqī writes: "Adam descended with a hollow red ruby that has four white corners *(arkān)* and laid it upon the foundation. It remained like that until the time of the deluge."[97] Al-Ṭabarī (d. 922) adds that the foundations of this house, which were laid out by the angels in the seventh earth, were also quadrangular.[98] The form of Adam's house was imagined as reflecting the form of the Frequented House, the Kaʿba's celestial model, described as "a building on four pillars of chrysolite that God crowned with a red ruby and called '*al-ḍurāḥ*.'"[99] Both Adam's house and its celestial model, *al-ḍurāḥ*, confirm the divine paradigm of the Kaʿba's quaternary structure. Ibn ʿAbbās is reported to have said: "God created the Kaʿba and placed it on water, upon four pillars, two thousand years before he created the world. From underneath the house the earth was then spread out."[100] As the omphalos whence the earth emerged and was spread out, the Kaʿba extended its quadrature into the shape of the earth. East, west, south, and north were depicted as four sides of the earthly square.[101] The spatiality of the Kaʿba thus becomes a meteorograph for the rainfall and fertility for all parts of the earth. "When rain beats one of the sides of the Kaʿba," a tradition says, "fertility will be during the year on that side; when it beats all sides, fertility will reign on all sides."[102] In representing the first spatial deployment, the Kaʿba becomes a visible trace of the creative process through which unity proceeded into four-ness and a tectonic expression of the emergence of space and directionality from its maternal, nondirectional source, the point. Many iconographic depictions emphasize the Kaʿba's quadrature and extend it to the complex of the *ḥaram*, its architectural context.

In Sufi views, the Kaʿba's quadrature is a reminder of the four directions along which Satan approaches and corrupts man, and of the presence of the four *awtād*, the spiritual guardians of these directions. Seen as "mountains"—"Have we not made the earth an expanse, and the high hills bulwarks *(awtād)*?" (78:6–7)—the *awtād* assume the role of *stabilizing* the faith of the believers and ensuring the constant flow of God's grace and inspirations.[103] Through the four *awtād* God preserves Islam as the primordial religion, *al-dīn al-ḥanīf.* By one he preserves faith *(al-īmān)*, by the second he preserves sainthood *(al-wilāya)*, by the third he preserves prophecy *(al-nubuwwa)*, and by the fourth he preserves scripture *(al-risāla)*.[104] In their onerous tasks, the *awtād* are aided by four prophets and four archangels, with which the four corners of the Kaʿba are identified.[105] Ibn ʿArabī explains:

> Among them is one who corresponds to the heart of Adam, the other to the heart of Abraham, the third to the heart of Christ, and the fourth to heart of Muḥammad, peace be upon them. So among them is one who is supported by the spirituality of Seraphiel, the other by the spirituality of Michael, the third by the spirituality of Gabriel, and the fourth by the spirituality of Izraiel. Each *watad* has one corner of the house: the one who corresponds to the heart of Adam . . . has the Syrian corner, the one who corresponds to the heart of Abraham . . . has the Irāqī corner, the one who corresponds to the heart of Christ . . . has the Yamanite corner, and the one who corresponds to the heart of Muḥammad . . . has the corner of the black stone.[106]

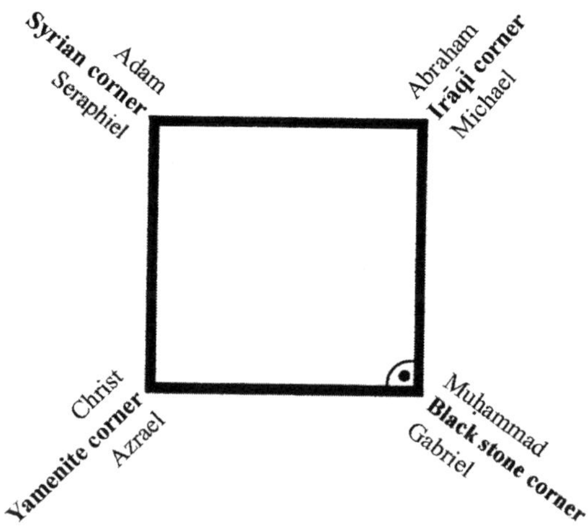

Fig. 4.12 The correlation of the Kaʿba's quadrature with prophet-angel pairing.

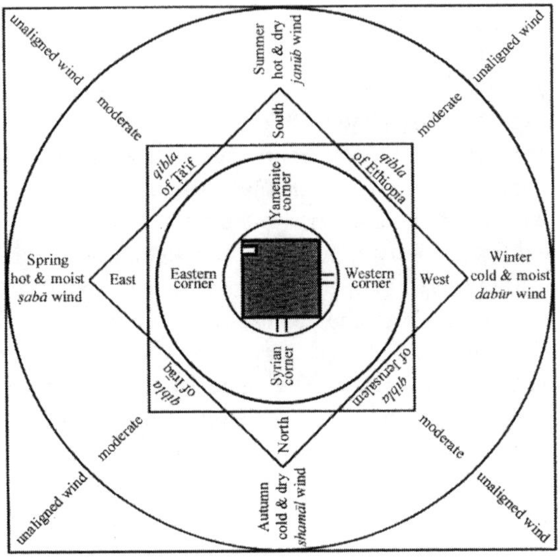

Fig. 4.13 The Kaʿba as a cosmograph of quaternary spatial structure (eighteenth-century Ottoman source cited in King, 1993, 8:308).

The corners of the Kaʿba correspond with the original divine quadrature of "the first and the last, and the outward and the inward" (57:3), which inheres in all created quadratures that God set for himself as the house of being.[107] In this correspondence, the inward correlates with the corner of the black stone, God's right hand on earth, which the pilgrims kiss in recognition of its significance. When the sight falls on the stone, Ibn ʿArabī explains, the insight falls on the right hand, the stone's inner reality. It stands for the "oil" of the "blessed olive tree," that is "neither of the east nor of the west," which sustains the divine light—"God is the light of the heavens and the earth" (24:35). While the outward form of the Kaʿba expresses directionality and spatial deployment, its centrality conceals the secret of the directionless *identity*, the *coincidentia oppositorum*, whence the light of the world emanates.[108]

In addition to its spatial symbolism, the Kaʿba and the rites associated with it also have temporal significance. A popular imagery depicts the Kaʿba with the circumambulating pilgrims as an earthly miniature of the divine Throne and its encircling angels. The imagery seemed so vivid that questions were raised about the nature of space occupied by the angels, since the Throne was known to have occupied the entire vacuum. And Ibn ʿArabī goes so far as to consider human glorifications of God in circumambulating the Kaʿba to be superior to those of the angels.[109]

The ritual circumambulation of the Kaʿba is performed in seven continuous, anticlockwise revolutions, starting from the corner of the black stone. A cycle is completed by the return to this same point, indicated by kissing, touching, or facing the stone when out of reach.[110] The black stone is seen to mark the starting point of the ritual revolution about the Kaʿba in the same way the position of the divine Feet on the Footstool regulates the cyclic revolution of the *aṭlas* sphere.[111] Thus the seven rounds performed about the Kaʿba correspond to the original cycles of the *aṭlas* sphere about the earth before the creation of the planetary spheres, and the delineation of the seven cycles by reference to the black stone reflects the differentiation of the seven divine days by reference to the divine Feet. It is with reference to this correspondence that the completion of the seven rounds around the Kaʿba is known as "completing one's week."

The seven revolutions of the *aṭlas* sphere are, as we have seen, temporal expressions of the seven principal divine attributes, and the ritual circumambulations are reenactments of the celestial cycles. Thus the seven cycles around the house correspond to the seven divine attributes, while the house stands for the Essence.[112] In realizing the significance of their act while partaking in the ritual circumambulation, the ambulants are said to be endowed with the qualities of the divine attributes.[113] By circumambulating the Kaʿba the ambulant not only reenacts the primeval process of temporal manifestation that successively took place through the revolution of the *aṭlas* sphere but also qualifies space by differentiating its four cardinal directions. The ritual circumambulation involves the utterance of four different statements that correspond to the four corners of the Kaʿba. These utterances punctuate the continuity of the ritual revolution, marking four distinct points that correspond to the four directions of space determined by the four corners of the Kaʿba. This act mimics the way in which the four nodal points—the two solstices and the two equinoxes—of the ecliptic punctuate the sun's annual journey whereby it measures out the limits of the space and defines the cardinal directions.

Triplicity and Quadrature

In *Akhbār Makka* Al-Azraqī relates a common *ḥadīth* that says that once the Prophet told his wife ʿĀʾisha that when her tribesmen of Quraysh rebuilt the Kaʿba, they did not lay it out exactly upon the foundations of Abraham. They made it shorter on one side by about seven cubits. Had they not been still well acquainted with infidelity, the Prophet lamented, he would have rebuilt the house, adding to it what they had omitted. The *ḥadīth* adds that the Prophet then took his wife and showed her the missing part of the house so that she might be a witness if Quraysh ever decided to add the missing part.[114] Acknowledging their omission, however, the Qurayshī builders fenced the unbuilt part of the Kaʿba by a semicircular parapet wall opposite the northwestern side. This fenced area is known as *al-ḥijir*, from *ḥajara*, literally "to deny access," "to detain." It was built in order to prevent the ambulants from intruding into that area during circumambulation, thus acknowledging its belonging to the Kaʿba.

At the time of Ibn al-Zubayr's revolt against the Umayyad caliph ʿAbd al-Malik, the Kaʿba was burned down. With reference to the above tradition, Ibn al-Zubayr is said to have rebuilt the Kaʿba according to the Prophet's descriptions exactly upon the foundations of Abraham. But no sooner had he finished

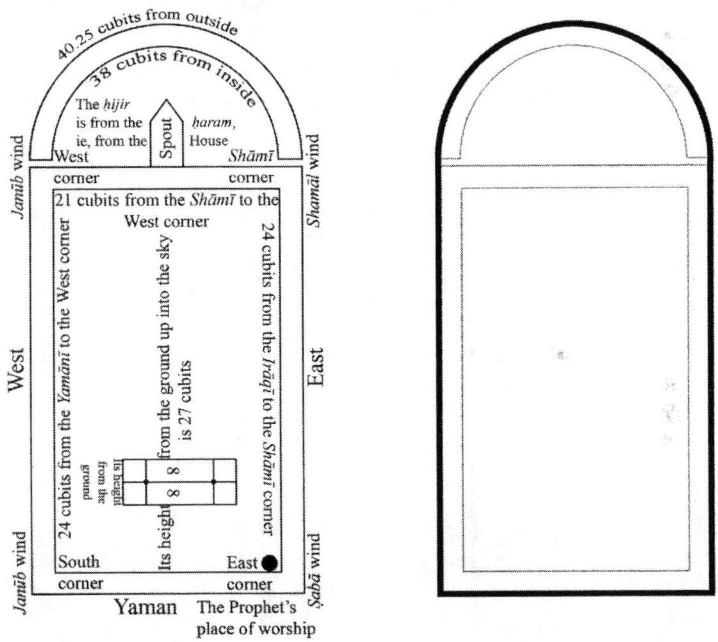

The form of the Kaʿba according to a medieval manuscript

The form of *al-ḍurāḥ* according to Ibn ʿArabī's *Futūḥāt*

Fig. 4.14 A plan of the Kaʿba showing the *ḥijir* (eighteenth-century Ottoman source cited in King, 1993, 8:308), and the form of the celestial archetype, *al-ḍurāḥ*, according to Ibn ʿArabī *(Futūḥāt)*.

rebuilding the Kaʿba than he was killed by al-Ḥajjāj, the Umayyad governor, who took over Mecca, pulled down the Kaʿba, and rebuilt it again as it was at the time of the Prophet. Thus it remained to this day. After realizing the authenticity of the above *ḥadīth*, some sources say, the Umayyad caliph ʿAbd al-Malik later regretted what he had done to the house and wished he had left it as it was.

Given the Kaʿba's rich cosmological significance, the above tradition and the related historical events raise critical questions about its original form and the divine model it meant to embody. Was it quadrangular or triangular? Ibn ʿArabī considers the Kaʿba to have two complementary forms, explicit and implicit. The explicit form is the current one with four corners *(arkān)*, while the implicit form is the one established by Adam and Abraham according to the divine model that has three corners only. He says that it was for a divine secret that "God caused his house to have four corners, though in reality it had only three, because it is *mukaʿʿab* in shape."[115] Here the term *mukaʿʿab* used by Ibn ʿArabī to describe the original shape of the Kaʿba calls for some explanation.

Although the Arabic term *kaʿba* denotes the idea of cube-like, it does not literally mean "cube," as is commonly understood,[116] nor does the Arabic term *mukaʿʿab*, currently used for "cube," refer, in the original sense of the word,

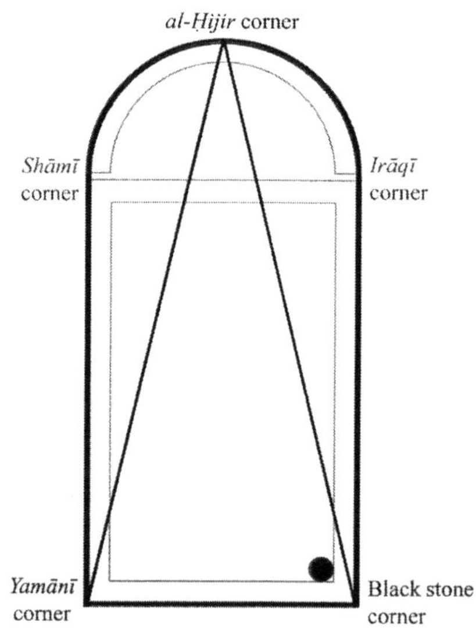

Fig. 4.15 Triplicity and quadrature as expressed in the implicit and explicit forms of the Kaʿba according to Ibn ʿArabī.

to a "geometrical hexahedron." This is evident from the preceding reference by Ibn ʿArabī. Both terms *kaʿba* and *mukaʿʿab* derive from *kaʿb*, which means literally "ankle," "anklebone," and "every articulation between bones." It seems that the form of the Kaʿba as a cube has something to do with *kiʿāb* or *kuʿūb* (plural of *kaʿb*), "bones," perhaps "cuboid," which were used in certain games in ancient Arabia.[117] The shape of these bones must bear some similarity to the semicircular form of the *ḥijir*, since Ibn ʿArabī says it is *mukaʿʿab* in shape. Describing the third corner of the Kaʿba's implicit from, Ibn ʿArabī says: "The one corner that coincides with the *ḥijir* is as the *ḥijir* in form, *mukaʿʿab* in shape, and for this reason it is called '*kaʿba*,' comparing it with the *kaʿb*."[118] The relationship between the circular form of the *ḥijir* and the name *kaʿba* can also be traced in the verb *kaʿaba*, meaning "to be full and round," "to swell," which was traditionally used to describe the "swelling of a woman's breast." Al-Zamakhsharī says that the "swelling" *(takaʿʿaba)* of a woman's breast is to protrude like the *kaʿb*.[119] Al-Farāhīdī further explains that *kaʿba* also means "quadrature," and the *kaʿba* of the house is its "upper quadrature," and every square house is *kaʿba*.[120] These references suggest that the term *kaʿba* refers originally to the form of the house including the *ḥijir*, that is, its quadrature as well as its semicircular end, rather than to the form of the cube, to which it nonetheless bears a secondary relation.

Be that as it may, Ibn ʿArabī considers the original divine model of the Kaʿba to be ternary rather than quaternary. The three corners of the house being the black stone, the Yemenite, and "the third corner is in the *ḥijir*." This corner is undetermined, he says, for it has no sensible form in the house.[121] As for the Syrian and Iraqi corners the *ḥijir* replaces, Ibn ʿArabī says that they were not part of the original divine model of the house and, consequently, not corners of the house, and this is the reason why their treatment in the pilgrimage rituals is different.[122]

The meanings of the Kaʿba's implicit triplicity, Ibn ʿArabī explains, lie in the holy tradition that is widely reported in Sufi literature, in which God says: "Neither my earth nor my heaven can encompass me, yet the heart of my faithful worshiper can." Sufis compare the heart to the Kaʿba for both share the notion of centrality at the micro and macrocosmic levels. Just as man is defined by two extremities, namely, his innermost heart and outermost body, so the world is defined by the Kaʿba, its innermost center, and the divine Throne, its outermost body. In a mystical reflection, Ibn ʿArabī poetically writes what God has communicated to him: "This Kaʿba of mine is the heart of existence, and to this heart my Throne is a defined body. Neither of them has space for me, nor tells about me what I tell about them, but the house of mine that has room for me is your intended heart, placed in your perceived body."[123]

In this analogical frame, the ambulants who frequent the Kaʿba become the "quick-passing thoughts" *(khawāṭir)* that frequent the heart. These *khawāṭir* are man's inner secrets *(asrār)* and interiorized thoughts that are generated by worldly engagements and personal preoccupations. They circumambulate the heart in the same way the pilgrims circumambulate the Kaʿba, so one's sensible attachments turn around one's body in the same way the angels move about the encompassing Throne. The ambulants of the Kaʿba are viewed to correspond to the secrets of the heart, because both share the idea of "heart-ness" *(al-qalbiyya)*; whereas the ambulants of the body correspond to the ambulants of the Throne, because both share the idea of "encompassed-ness" *(al-iḥaṭiyya)*. And just as the composed body is of a lesser status than the simple heart, so likewise is the encompassing Throne in relation to the Kaʿba.[124]

As for the quick-passing thoughts *(khawāṭir)* that circumambulate the heart, Ibn ʿArabī says, they derive from four distinct sources: divine, angelic, psychic, and satanic. They correspond with both the quaternary and the ternary orders of the Kaʿba. When we consider the ternary order, Ibn ʿArabī says, the three corners of the black stone, the Yemenite, and the *ḥijir* correspond to the heart's divine thought *(al-khāṭir al-ilāhī)*, angelic thought *(al-khāṭir al-malakī)*, and psychic thought *(al-khāṭir al-nafsī)* respectively, leaving no room for the satanic thought *(al-khāṭir al-shayṭānī)*. This is the model after which the hearts of prophets are made, he says, and to which Satan has no access.[125] But when we consider its quaternary pattern, the four corners, with the Syrian and the Iraqi corners replacing the *ḥijir*, we add the satanic thought. From the ritual invocation associated with each corner during circumambulation, Ibn ʿArabī infers that the Irāqī corner is the one correlated with the satanic thought, whereas the Shāmī corner is the one correlated with psychic thought. The quaternary order underlies the model after which the hearts of all people, other than the prophets, are made, including the hearts of the faithful.[126]

Concerning the relationship between the concealed triplicity and revealed quadrature, Ibn ʿArabī says that it was the divine wisdom that prompted the Umayyad caliph to order his governor al-Ḥajjāj to pull down the Kaʿba and return it to the current quadrant form that does not conform to the original divine model.[127] In this the Kaʿba become a true expression of the creative relationship between hidden triplicity and manifest quadrature, the triplicity of the divine command *(al-amr al-ilāhī)* and the quadrature of universal manifestation *(al-ẓuhūr al-kullī)*. The concealed triplicity complements the revealed quadrature in the form of the Kaʿba, in the same way in which spirituality complements materiality in cosmogony, and essentiality complements substantiality in manifestation. Thus this productive relationship that was first revealed through the inherent nature of the intermediary world of *al-barzakh*, and then manifested in various modes at different cosmic and human levels, is ultimately crystallized in the architectural form of the Kaʿba.

The Mosque and the Spatiality of Prayer

Mosques—*masjid*, *muṣallā*, and *jāmiʿ*—are the Islamic places of worship the Muslims have constructed in a rich variety of forms and styles. Austere or elaborate, simple or monumental, the mosque serves a uniquely Islamic function and has, therefore, been widely recognized as the expression par excellence of Islamic architecture.[128] Despite the large body of literature on the origin and development of mosque architecture, the curious relationship between the act of prayer and the architecture of the mosque, between the function and the form, has rarely been profoundly explored. Several prophetic sayings, such as, "Wherever you pray, that place is a mosque," and, "I have been given the whole earth as a sanctuary," raise questions about how and why an identifiable mosque architecture emerged and developed. In its formal and compositional characteristics, the typical mosque remains an intriguing phenomenon that is at once simple and complex. It is simple in that a number of recurrent elements can be traced in various compositions throughout the premodern and modern periods, revealing a consistent identity. Yet it is complex in that the model perpetuated in many elaborate forms has little to do with the function it serves. One only needs to observe the ranks of a large Muslim congregation during a major feast or Friday prayer, extending in linear form inside and outside the prayer hall, inside and outside the courtyard (when a mosque has one), to realize how impertinent the form of the mosque is, with all its elaborate elements, to the fulfillment of the prayer. If one can pray just as well inside and outside a mosque, then what difference does architecture make? This question is further complicated when one considers the various curious extensions, enlargements, and additions made to many mosques, such as, for example, the duplication of al-Manṣūr's mosque and the successive additions to the Cordoba mosque; or considers the large structures that are sometimes built within the prayer hall, as with the shrine of the prophet Yaḥyā in the great Umayyad mosque of Damascus; or considers the broad range of structures and forms to which the term "mosque" can apply. For instance, the Dome of the Rock, the sanctuary where it stands, and the Aqṣa are all referred to as "mosques" in premodern Islamic literature. With regard to the spatial order traced so far, the typical hypostyle mosque layout appears to bare little relevance. In order to show the consistency of the premodern spatial sensibility with regard to mosque architecture, I have followed an unorthodox approach that focuses on the spatiality of prayer.

The Prophetic Model

Historians of Islamic architecture generally concur that the simple dwelling and hypostyle mosque constructed by the Prophet in Medina served as a model for mosque design throughout the Islamic world.[129] Some even argue that this model contained in an embryonic form the elements that later became the defining characteristics of mosque architecture. The perpetuation of the basic layout of the prophet's house-mosque is curious, since this prototype is not conceived as being constructed according to a specific divine model. Unlike the Kaʿba, the first mosque did not generate a rich body of mythical narratives associated with its founding and construction.[130] This may be explained by the lack of Quranic references and the nature of prophethood in Islam; however, it seems that the prophetic example has in time assumed the status of a divine model that legitimated its numerous adaptations.

The story of founding and constructing the Prophet's house, however, is not without sacred connotations. Early sources describe how the Prophet was led to the site of his mosque by God.[131] As he entered the Medina on the back of his camel, al-Qaṣwāʾ, Muḥammad was fervently welcomed, and many eager invitations were offered. But Muḥammad told the inviting crowd repeatedly to let his camel go on her way saying: "She is under the command of God" *(maʾmūra)*. After a short tour, al-Qaṣwāʾ finally knelt and flattened her chest against the ground. There, Muḥammad alighted and said: "This, if God will, is the dwelling."[132] The sources also report that during the course of construction, Muḥammad asked his companions who were helping him to "build it in the form of *ʿarīsh* like that of Moses,"[133] indicating that he had a particular model in mind. According to the literal meanings of the term, *ʿarīsh* refers to a "trellis" erected for shading, to the roof of a building, probably made in the form of a trellis, to a kind of house found in Mecca, and also to "tabernacle" or "tent."[134]

The Prophet's house-mosque was in the shape of a square measuring one hundred or seventy cubits a side. It consisted of two parts: a simple walled courtyard and a simple shelter built with a flat roof held up by palm trunks used as pillars. This was first built along the northern side that faces Jerusalem, and a few months later with the divine injunction to change the *qibla* it was dismantled and reerected along the southern side, which faces Mecca. This shelter formed the prayer hall and opened onto the courtyard, which formed an extension to the covered area and an integral part of the mosque as a whole. The open courtyard was the predominant part of the mosque and was provided with three doors, one on each of the eastern, western, and northern sides of the square.[135] In his *Ṭabaqāt*, Ibn Saʿd (d. 845) describes the doors as being one at the rear, one the door of *al-Raḥma* (compassion) that is also known as the door of *ʿĀtika*, and the third was the door through which the Prophet used to enter.[136] Whether these three doors together with the sheltered hall, often depicted in opposite positions, were meant to mark the four directions of space is not clear. Yet, it is clear that many subsequent mosques, including the early great Umayyad mosque of Damascus that was first to reproduce the Prophet's model at a monumental scale, reveal similar planning and spatial characteristics. The question that is rarely discussed in the studies of mosque architecture is how does this model serve the performance of prayer?

Prayer as Visualization

As prescribed by the Prophet, Islamic prayer is an act of worship performed toward a liturgical center, the Kaʿba, that, unlike the church's alter, lies beyond the boundaries of all mosques, except for the one that contains the Kaʿba. The Prophet further teaches: "adore God as though you do see him, for if you do not see him he does see you." Thus viewed, the Islamic prayer is not primarily a pictorial experience or a visually oriented act, for the object of *seeing* is that which cannot be seen. In the Islamic prayer "seeing" takes on a different meaning, especially when viewed from the Sufi perspective. Islamic prayer requires no tangible object, such as an icon or a statue, to induce a sense of divine presence and serve as a support for worship. Visual engagement is therefore unnecessary. The only visual engagement it requires is that whereby Muslims orient themselves toward the *qibla*. Otherwise, the prayer is simply a bodily performance associated with oral recitations, requiring, especially in the communal prayer, an acute auditory engagement. From the moment the call to prayer is heard, Muslims engage in aural-oral correspondence, repeating certain phrases and acting in certain ways. During prayer, the oral recitations by the *imām* are the principal means of regulating the prayer's rhythmical sequence. This was the reason why in some large mosques the *dikka*, a "respondents' platform," became a necessary piece of liturgical furniture in order to extend the *imām*ʾs audible presence.

There are several prophetic traditions that define the nature of the Islamic prayer, such as the one cited above and the one that says: "In truth God is present in the *qibla* of every one of you." But the most important one in the context of this study is the holy tradition that concerns the recitation of *al-fātiḥa*, the opening chapter, which constitutes the principal text of the prayer. In this rather long *ḥadīth*, God begins by saying: "I have divided prayer between me and my servant into two halves, one being due to me, the other to my servant; and my servant will receive that for which he asks."[137] The *ḥadīth* goes on depicting the prayer as a dialogical act, a direct conversation between God and Man, wherein each has a role to play and a responsibility to fulfill. In this sense, prayer becomes as it were a "colloquy" between the adorer and the adored, an "intimate dialogue" *(munājāt)* between the creator and his creature.[138] Ibn ʿArabī goes so far to suggest that prayer is indeed a shared act of worship: "He glorifies me

Photo 4.4 The respondents' platform *(dikka)* at the Sultan Ḥasan school in Cairo.

and I glorify him. He worships me and I worship him."[139] In *Mir'āt al-ʿĀrifīn*, the author translates this holy tradition into a diagram that illustrates the dialogical nature of prayer in a geometrical form.[140]

Ibn ʿArabī teaches that to achieve such a dialogue, one must place oneself in the presence of God, imagine God as being present in the *qibla*, visualize his presence in his heart, and hear his voice vibrating in all manifested things.[141] *Shuhūd*, "vision" or "visualization," the key Quranic concept in such experience, refers to *imaginative visualization* that compensates for the lack of sensory visual engagement in prayer. Imagination being our only means of engagement with the Absolute, any prayer can be seen as an attempt to communicate with God through imaginative visualization. The way to achieve such visualization, however, may differ. In Christianity, for example, Ibn ʿArabī observes that visualization is achieved through the mediation of icons. "The Byzantines developed the art of painting to its perfection," he writes, "because for them the unique nature *(fardāniyyah)* of Sayyidnā ʿIsā as expressed in his image, is the foremost support of concentration on Divine Unity."[142] In contrast, Islam prohibits the use of icons in prayer, he adds, prescribing instead to "adore God as though you do see him." Ibn ʿArabī's argument reveals an awareness that visualization in Christianity is induced through the agency of

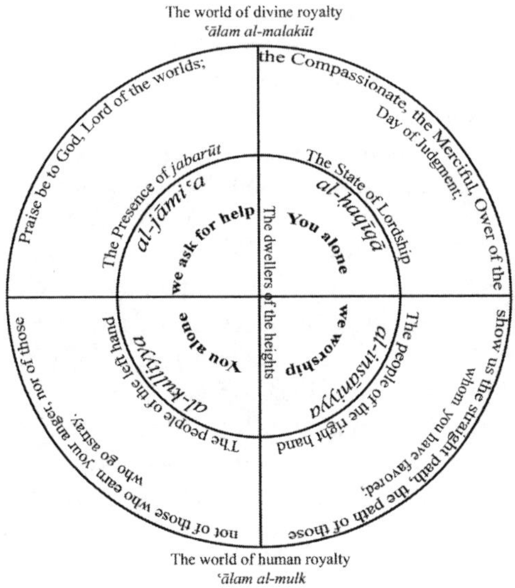

Fig. 4.16 The geometry of *al-fātiḥa* ("Mirʾāt," MS. 4865).

depiction; whereas in Islam it is induced through the agency of *invocation*. Depiction centers on the divine word revealed in the form of a human being, as in the case of Christ, demanding direct visual engagement with the depicted *image*. By contrast, invocation centers on the divine word revealed in the form of a book, as is the case of the Quran, demanding direct auditory engagement with the invoked *text*.

On a practical level, Muslim clerics have considered some techniques for concentration that would enhance the experience of praying. Since closing the eyes in prayer is undesirable in Islam, they searched for ways to reduce visual distractions.[143] In the *Iḥyāʾ*, for example, al-Ghazālī advises that one should restrict one's sight to the spot on which one is standing, which can be achieved by standing near a wall or by drawing a line on which one's gaze is fixed throughout the prayer. In doing so, he says, one shortens the range of one's sight and consequently enhances one's concentration.[144] Whereas vision is regarded as distracting in prayer, hearing is considered to be engaging. In any case, one cannot do away with one's hearing in the same way one can with one's vision by simply closing one's eyes.

Sufis associate hearing with invocation *(dhikr)*, an act that creates a sense of divine presence necessary for prayer. "Whoever invokes God, finds himself in the presence of God," Ibn ʿArabī writes, referring to the holy tradition that says: "I witness the invocation of he who invokes me."[145] *Dhikr* means "recall" and "remembrance." When identified with prayer, *dhikr* becomes the recalling and remembering of God in order to converse with him. Having the ability to invoke is only one side that must be complemented by the ability to "hear" the divine reply. If one is unable to hear God's reply in the prayer, Ibn ʿArabī stresses, then one is not "lending his hearing while witnessing" (50:37); that is, one is not present in front of one's Lord and can neither hear nor see him. Thus understood, it is not the eye and the sense of seeing that enhance the creative imagination in prayer, but rather the ear and the sense of hearing. Hearing engenders the experience of a presence, and the oral-aural participation activates the inner sense of vision, enabling one's visualization. "He who makes himself present to whom he invokes, being one with in-sight, *sees* his companion," Ibn ʿArabī explains, "this being both a visualization *(mushāhada)* and vision *(ruʾya)*." If one cannot attain such visualization, however, one should worship God by faith as though he sees him, that is, imagining him in his *qibla* while conversing with him and lending his hearing to what he might receive from him.[146]

Prayer and Acoustic Space

Islamic prayer thus understood does not depend on the eye and the sense of seeing and, therefore, bears poor relationship to pictorial or visual space. Rather, it depends on the ear and the sense of hearing, bearing a stronger affinity with auditory or acoustic space.[147] The main characteristics of acoustic space derive from sound's essential feature of invoking an unlocalized presence. "Because of its association with sound," Ong observes, "acoustic space implies presence far more than does visual space."[148] Sound suggests a presence without location, a presence that occupies the entire space rather than being located in it. This is very much like God's presence in the *qibla* of the faithful, which, though physically in front of him, is in reality everywhere around him: "Wheresoever you turn, there is the face of God" (2:115).

Unlike visual space, auditory space is not spread out before us with a fore, middle, and background but is diffused around us as a boundless bubble, as a sphere that has no precisely defined boundaries.[149] Sound can be heard equally well from any direction, front or back, left or right, above or below, and in any position, lying down, sitting, or standing up. Accordingly, the space apprehended by the ear has neutral spatial characteristics: neither does it have any favorable point as a center, nor does it relate to any particular direction more than others. Every point in it is a center sufficient to itself, and every spot entertains the sense of spatial entirety. In an acoustic space every participant is situated in the center of his or her own acoustic field, regardless of the location or the number of participants. Furthermore, auditory space is not a pure or empty, boxed-in space but is an "essentially inhabited space."[150] This means that one cannot simply be independent of it; one cannot stand outside it and experience it but has to be within it in order for such space to exist. Acoustic space cannot be frozen in the matrix of matter; it cannot be preserved, as in a photographic form, and made to passively endure in the memory. For there is no way to preserve sound as sound; when sound stops, its opposite, silence, prevails.[151] Accordingly, acoustic space is not a passive, static space in the sense that its inhabitant adds virtually nothing to its quality; rather, it is an active, dynamic space, "always in flux, creating its own dimensions moment by moment"[152] and always in transformations, corresponding to the states that one may attain during one's presence in it.

The characteristics of acoustic space provide an ideal environment for the performance of Islamic prayer, in which every participant is invested with the dignity of the *imām* and acts as his own priest. A reported prophetic tradition says that every worshiper is an *imām*, for the angels pray behind him when he prays alone.[153] Theoretically, every Muslim praying constitutes an independent

Photo 4.5 The interior of the great Umayyad mosque of Damascus.

Photo 4.6 The tomb of the prophet Yaḥyā inside the prayer hall of the great Umayyad mosque of Damascus.

center directly connected to the supreme center, the Kaʿba, in the same way that every participant in an acoustic field finds him- or herself directly related to the sonic source. Although architecture can enhance the acoustic qualities, auditory space is not a tectonic space. It is a space that can virtually exist anywhere, as a prophetic tradition asserts: "God has blessed my community by giving them the face of the whole world as a sanctuary."[154] But Muḥammad did not pray anywhere. As we have seen, he did project a choice.

Examining the spatial characteristics of a prayer hall of the early hypostyle model from the standpoint of acoustic space, the following spatial characteristics emerge: short visual field, no visual center, and neutral spatial characteristics. The most conspicuous quality one observes is that the space does not beckon the eye in a particular direction. Looking in a direction parallel to any row of columns, the eye neither rests on a terminating element nor is invited to any point of special significance. Looking diagonally, however, one encounters a forest of columns, rendering the visual field very limited. Mosque spatiality lacks orientation toward a liturgical center comparable to the one found in a church, which acts as both a spatial and a visual focus.[155] It includes no starting or terminating point that may evoke a sense of progression. As a result, a space of this kind can be extended in practically any direction, as had actually happened in many mosques; it can accommodate varying ranks of pillars without affecting its internal spatial order or changing its essential characteristics. "In this respect," Grabar observes, "the early mosque was a remarkably modern building which could be expanded and contracted according to the needs of the community."[156] A prayer hall can also include large structures, such as the large tomb of the prophet Yaḥyā in the great Umayyad mosque of Damascus, without disturbing the sense of spatial cohesion and continuity.

Thus viewed, the spatial characteristics of the prayer hall of the early hypostyle mosques provide a suitable ambience for praying, through the sense of neutrality, nonprocession, repose, and equilibrium it reveals, allowing every point in the space to be a center of equal significance.[157] Such an ambience tends to generate an encompassing sense of presence that is consistent with the Islamic idea of universal sanctuary.[158] Through oral-aural participation in prayer, this ambience helps evoke one's own inward auditory space, which is "in a way a vast interior in the center of which the listener finds himself together with his interlocutors."[159] This, however, should not be taken to suggest that this was intentional in the prophetic model. Yet the indifference the function of praying presents toward architectural forms might be seen as having helped the adoption, adaptation, and perpetuation of a prophetic model in wherever form it happened to be.

The Spatiality of Prayer

Acoustic space, as we have seen, cannot be dissociated from the individual who is the center of this space. The experiences and meanings associated with this space, therefore, derive from the actions of the one who at once unfolds and occupies this space. In the course of prayer, each individual defines a spatial field by his or her bodily movements that have spiritual significance. It is, therefore,

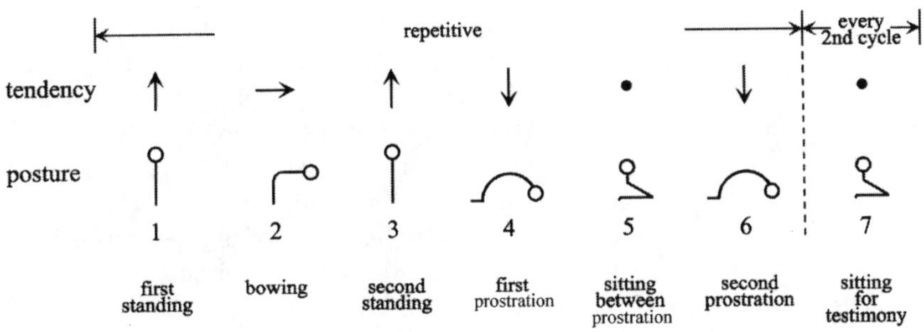

Fig. 4.17 The bodily postures and associated spatial tendencies of the Islamic prayer.

important to know the bodily and oral performances involved in prayer as well as their sequence to fully appreciate their spiritual significance.

Islamic prayer comprises a prescribed set of gestures and recitations, performed in the same way individually or collectively while standing at a fixed point in space. It involves a series of bodily postures rhythmically repeated in one place with no processional rituals. There are four principal postures: standing *(qiyām)*, bowing *(rukūʿ)*, sitting or resting *(julūs)*, and prostrating *(sujūd)*. The movements associated with these bodily postures reveal four tendencies: upward, associated with the standing posture; horizontal, associated with the bowing posture; downward, associated with the prostrating posture; and stillness, associated with the resting posture. When repeated in a certain sequence in association with an oral recitation, these postures constitute the so-called *rakʿa* (from *rukūʿ*, "bowing posture"), the prayer's repeated unit or cycle. Each complete *rakʿa* (cycle) consists of seven distinct acts: six repetitive ones and a terminating one every two cycles, as shown in the diagram. When a prayer is of three or five cycles, the seventh act is also repeated in the last odd cycle.

Sufis see in the spatial tendencies of the prayer an expression of the three-dimensional cross, the underlying divine structure of both human and cosmic formation, and the basis of spatial ordering. Ibn ʿArabī explains: "Since Being became known through an intelligible movement that transferred the world from nonexistence to existence, the prayer involved all the movements, which are three: the rectilinear movement which is the worshiper's standing posture, the horizontal movement which is the worshiper's bowing posture, and the reversed movement which is the worshiper's prostration. Man's movement is rectilinear, the animal's horizontal, and the plant's reversed; the mineral has no movement of its own, when a stone moves it is being moved by other forces."[160]

Thus viewed, the three movements of the prayer reenact the primordial process of existential unfolding, which, according to al-Qāshānī, occurred by way of three intelligible or cosmic movements. They also retrace both the movements of spatial expansion from the center and the natural growth of the human body from the sacrum. Accordingly, by the tendencies of their bodily movements, a Muslim in prayer unfolds—in principle—a sphere that defines the spatiality of the boundless bubble of their acoustic field. Ibn ʿArabī describes an experience he once had in a prayer wherein he visualized himself as being transformed into intense light wherein he could no longer distinguish his directions.[161]

The postures of the prayer together with their associated tendencies are seen to correspond to the threefold origin of humanity. Through the prostrating posture and its downward tendency humans seek the origin of their body, water and earth; through the standing posture and its ascending tendency they seek

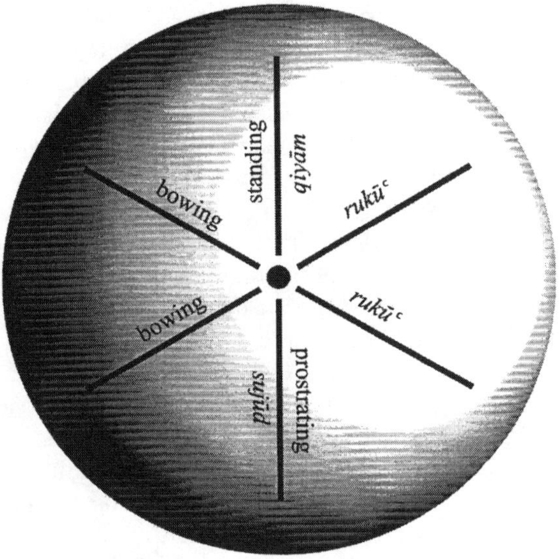

Fig. 4.18 The spatiality of Islamic prayer as it unfolds in the bodily act of praying.

the origin of their spirit, God; and through his bowing posture and its horizontal tendency they seek the origin of their imaginative and mental faculties, the isthmus *(al-barzakh)*.[162] The horizontal, bowing posture mediates between standing and prostrating and thus includes in its form the qualities of both: ascending spirituality and descending materiality. This isthmian characteristic, Ibn ᶜArabī explains, is expressed though the phrase the Muslim utters on God's behalf while rising from the bowing position: "God hears him who glorifies him." The bowing posture also corresponds to the isthmian verse of *al-fātiḥa:* "It is you whom we adore, and it is of you that we beg assistance," which God and man share and which divides the seven verses of *al-fātiḥa* into two halves: three relating to God and three relating to man. The postures of prayer are also viewed to correspond to the form of the Arabic letters. The standing posture corresponds to the letter *alif* (A), whose verticality is seen to represent the original *position* from which other postures derive. Calligraphically, the straightness of the *alif* is taken to represent the principal form of the various curvatures and bending of other letters.

Considering the ritual prayer during the pilgrimage, Ibn ᶜArabī constructs another level of cosmological significance based on the spatiality of prayer.

The pilgrimage rituals prescribe the performance of two *rakʿas* after the completion of every seven laps around the Kaʿba. The main reason for this, Ibn ʿArabī explains, comes from the very act of circumambulation. In circumambulating the Kaʿba seven times one assumes the role of the revolving planets in generating and ruling over the earthly conditions. The planets are cosmic agents with which God regulates the worldly affairs through their influences on the four natural elements. In reenacting the "days" of the divine "week," the ritual circumambulation assumes the function of the celestial agents, and the human body assumes the place of the four elements. This cosmological resonance necessitates the two *rakʿas*, since the human formation has two components, material and intellectual. "God also rendered for every cyclical movement of this week a trace in the prayer," Ibn ʿArabī further explains. "In the prayer there manifests seven bodily and spiritual traces, one trace from the movement of every cycle of the week of circumambulation . . . The seven bodily traces in the formation of the prayer are first standing, bowing, second standing, which is rising up from bowing, [first] prostration, sitting between the two prostrations, second prostration, and sitting for testimony. As for the invocations associated with these seven bodily movements, they are their spirits. Thus payer was composed as a perfect formation."[163]

Unity and Community

In communal prayers individuals are required to be perfectly aligned in straight lines with shoulders rubbing so that no spaces are left between the participants. The perfect alignment is seen as a spatial expression of the equality all worshippers have to God. God asks humans to pray in order to *commune* with them as a community. As earthly creatures bound to space, humans express their equal connection to divinity spatially though their perfect alignment, which images the way in which angels stand in God's presence: "And your Lord shall come with angels, rank on rank" (89:22), "on the day when the angels and the Spirit stand arrayed" (78:38). The *imām*, however, leads the prayer from his stand-alone position. Here the spatiality of the *imām* and the aligned crowd involve a double resonance.

In standing alone, Ibn ʿArabī explains, the *imām* represents the totality of the worshipers, constituting a row on his own. No one else is allowed to do so. If one finds oneself alone one is required to either squeeze in the last row, or, if it is already well-packed, invite one from the row to make a pair. Standing alone, the *imām*'s prayer projects God's "unity of totality" *(aḥadiyyat al-majmūʿ)*, whereas standing in aligned rows, the prayers of individuals project God's "totality of unity" *(majmūʿ al-aḥadiyya)*. The inversed resonance reveals an awareness of the spatiality of visualizing and projecting divinity by the *imām* and the community in the course of prayer. The spatial design of the prayer hall of the hypostyle mosque seems to reflect this reciprocal relationship. The pillared hall tends to express architecturally the congregation's "totality of unity," whereas the *imām*'s position marked by the *miḥrāb* and sometimes a dome positioned directly above it, or even the central dome, tends to express the *imām*'s "unity of totality."

Creative Hearing

Finally, hearing *(samāʿ)* recalls the creative act of divine utterance and the process of existentiation *(ījād)*. In Ibn ʿArabī's cosmology, *utterance* and *hearing* are the two generative principles of existentiation. Every determining agent is a father, every determined thing is a mother, every act of determination is a marriage relationship, and every resultant is a child. The speaker is a father, the listener is a mother, the act of speaking is a marriage relationship, and the understanding or the vision produced in the listener is a child.[164] The Quran often describes God as "the Hearer, the Knower" and "the Hearer, the Seer." Hearing is associated with Knowing and Seeing, but hearing always comes first. Ibn ʿArabī says that this is because hearing was the means by which existents first knew of their Lord when they responded to the creative command "Be!"[165] Hearing, as we have seen, was also the first faculty possible beings received concurrently with the first temporal motion, engendered by the attribute of Hearing, whereby the first day of the divine week occurred. The "hearing" of possible beings is what brought them into existence and enabled them to then see and know their creator.[166]

Hearing, Ibn ʿArabī explains, is of three kinds: divine hearing *(samāʿ ilāhī)*, spiritual hearing *(samāʿ rūḥānī)*, and natural hearing *(samāʿ ṭabīʿī)*. Divine hearing is the unconditioned hearing *(al-samāʿ al-muṭlaq)* of the "ears of the heart" that relates to the transcendental sound of the divine creative command heard from everything, in everything, and with everything, as expressed in the verse: "Never comes there unto them a new reminder from their Lord but they listen" (21:2). It is the hearing of one for whom existence is the inexhaustible divine words, and of whom God says: "I am the hearing wherewith he hears." Spiritual hearing is the cosmic hearing *(al-samāʿ al-kawnī)* of the "ears of the intellect" that relates to the scratching sound *(ṣarīf)* of the supernal Pen as it inscribes the divine words upon the Preserved Tablet of existence. It is the hearing of one who sees the traces of the divine words in the cosmic forms. Natural hearing is the hearing of the "ears of carnal soul" that relates to natural sounds.

Natural hearing reflects spiritual hearing, which in turn reflects divine hearing. They are all based on the pattern of quadrature. Divine hearing involves the Essence *(dhāt)*, relation *(nisba)* or attribute, turning toward the listener *(tawajjuh)* or will, and utterance *(qawl)* or power. Spiritual hearing involves the Essence *(dhāt)*, hand *(yad)*, pen *(qalam)*, and the scratching sound of inscription *(ṣarīf)*. And natural hearing involves the four principles of nature: heat, cold, moistness, and dryness. By the quadrature of divine hearing Being *(wujūd)* manifests; by the quadrature of spiritual hearing the Universal Soul manifests, and by the quadrature of natural hearing natural existence, to which the human sensible body belongs, manifests.[167] Ibn ʿArabī says that those who belong among the people of the divine hearing observe the metaphysical order of the divine names; their hearing and consequently their knowledge derive from there. Those who belong among the people of the spiritual hearing observe the traces of the metaphysical order in the cosmic order of both the higher and lower worlds; their hearing and consequently their knowledge derive from there. As for those who belong among the people of natural hearing, they can only feel and respond to the effect of natural sounds; natural hearing does not produce knowledge; it only produces an effect in the soul.

Afterword: Architecture and Cosmic Habitat

Al-Ghazālī's analogy that has guided us through the preceding journey conceals hitherto unexamined assumptions concerning the act of designing. First is the predictability of God's way of designing and our ability to read God's mind by reflecting on his design work. Second is the architect's responsibility to emulate the predictable aspects of God's design. The analogy has led us to explore what it could possibly mean to consider the question of design as an ontological enquiry and to view it from a metaphysical perspective. The exploration revealed what set of existential possibilities the cosmogonic act of design can bring together, what levels of understanding designers can attain when they see themselves as privileged creatures strategically positioned at the center of the universe, and what it means for these creatures to see themselves as the center, the model, and the ultimate aim of existence. With and without the mystical overtone, the awareness of the centrality of humankind in a universe that is purpose-designed for their dwelling and habitation is as relevant to us today as it ever was.

At the turn of the twenty-first century certain fundamental questions concerning the nature, purpose, and order of existence have persisted. The premodern cosmological picture, however, has been irreversibly altered. Physical sciences aided by advanced technologies have probed new frontiers of reality at scales, distances, and complexities unimaginable before. Today, the cosmic terrains are remarkably vaster, the temporal scope is considerably deeper, and the cosmological scenario is far more complex. Yet modern and premodern conceptions seem to overlap in certain speculative areas concerned with the beginning of the cosmic drama. "Everything astronomers can see, stretching out to distances of 10 billion light-years," cosmologist Martin Rees asserts, "emerged from infinitesimal speck."[1] What followed is of course a different story; however, the *logic* of the beginning remains consistent. As to why anything exists at all and why things are the way they are remain in the realm of speculations outside the purview of physical sciences.

Modern cosmology has led to two significant changes: the decentering of humankind in the universal order of existence and the fragmenting of the spatial sensibility that underpins the meaningful continuity between everyday life and other levels of reality. More by default than intention, modern cosmological theories trivialize the role of humankind in the scheme of existence. In the mind of God, man is no longer the original idea. As Paul Davies observes, "People were no longer cast at the center of the great scheme, but were relegated to an incidental and seemingly pointless role in an indifferent cosmic drama, like unscripted extras that have accidentally stumbled onto a vast movie set."[2] Furthermore, our spatial sensibility is being constantly reshaped at the frontiers of scientific and technological inventions, at the cutting edges of "the very *big*, the very *small*, and the very *complex*."[3] Today, quantum and astrophysicists explore reality at either ends of a spatio-temporal scale that extends between 10^{-20} cm and 10^{+25} cm, dealing with unimaginable dimensions that may make sense mathematically but remain experientially incomprehensible.[4] What connects us with the cosmos from the scientific perspective is no longer a structural or ontological resonance but rather the very stuff we are made of, the chemistry of our bodies that "derived from stellar explosions that occurred well before the formation of the solar system."[5] We are simply *animated stardust*. This is where the existential continuity is currently pursued. As Rees observes, "Almost all scientists already believe that our everyday world is, in a sense, reducible to atomic physics."[6] By understanding the workings of the atomic and subatomic structures, it is argued, we will be able to understand the order and meaning of the universe. The theory of everything, which promises to give us the ability to read the mind of God, does not start with the whole but with the nature of the invisible units of being. As for the big picture, the deep forces that shape the universe are also pursued in mathematical terms that, like the theory of everything, bare little relevance to everyday life and unmediated human experiences. Reconciling the physics of the very small with that of the very big might be one of the most exciting scientific projects, but making sense of such reconciliation at the human scale remains the real challenge.

The issue of scale is important, particularly with regard to the question of cosmology and architecture. Science fiction might have already projected numerous scenarios of what it means to live and think at the intergalactic or atomic levels, but these hyper-real spaces remain alien to the daily activity, to the Euclidean space where we live, interact, design, and build. Premodern spatial sensibility was anchored in the Euclidean space whose order applied consistently from the big to the small ends of reality. Today, architecture while remaining faithful to the principles of Euclidean geometry, both in experience and modes of production, has been alienated from the curved galactic spatiality of relativity and the multidimensional subatomic spatiality of the superstrings. The architecture of the universe and earthly architecture are no longer encompassed by the same perspective. It is this alienation that broke the structural nexus between man and the cosmos, resulting in significant reshaping of our spatial sensibility. The formal and structural correspondences between man and the cosmos that lay at the heart of the philosophical order discussed in the preceding pages have now lost their *reality*; and the countless analogies among the human body, earth's geography, the cosmos, and the constructed environment have now lost their immediacy as objects of belief. The idea of structural resonance between the whole and the parts has also lost its currency in today's scientific enterprise.

Life and intelligence, however, remain the puzzling elements that compel scientists to ponder the exceptional "recipe" of existence and to believe in the uniqueness of our planet. Self-consciousness "can be no trivial detail, no minor by product of mindless, purposeless forces," Davies observes.[7] If "we are truly meant to be here," as he concludes, then Rees's idea of *cosmic habitat* assumes a special significance, for it offers a new possibility of crossing between architecture and cosmology. The growing global concern for the future of our planet and the sustainability of earthly habitats has already been appropriated into the mainstream architectural discourse. Rees's notion of cosmic habitat broadens our perspective to include the universality of our cosmic environment, prompting us to think not only of the earthly conditions but also of the very fine universal balance that underpins and sustains the presence of life. Moreover, the notion of cosmic habitat, while expanding our perspective of the "native" habitat, has the potential of bringing back humankind into the central stage of existence. It introduces a new dimension, new environmental scale, where the "native" environment for habitation extends beyond the confines of regional geography: the "native" becomes the "global" and its cosmic conditions. This brings about new challenges to architectural theorists who have to consider the ontology of design and sustainment in the complexity of today's cosmology. It is an invitation to reinvent, as Karsten Harries puts it, "an altogether new postmodern geocentricism."[8]

Notes

Introduction

1. Al-Ghazālī, *Iḥyā'*, 3:20–21.

2. For example, see the listed works of Péres-Gómez, Wertheim, Foster, and Harries.

3. Many mystical ideas are also traceable in the writings of the twentieth-century Sufi Aḥmad al-ʿAlawī (d. 1934). This shows the consistency of the Sufi thought even beyond the medieval period. See Lings, 1973.

4. Guénon, 1975.

5. See Snodgrass, 1990.

1. Discursive Order

1. For a review of Islamic cosmology, see Karamustafa, 1992; Heinen, 1982; Nasr, 1978a; and Jachimowicz, 1975.

2. See Heinen, 1982.

3. On the *kalām* cosmology, see Craig, 1979; on the Muʿtzilites' approach, see Abū Zayd, 1996.

4. See al-Ghazālī, *Tahāfut*; Ibn Taymiyya, *Fatāwā*.

5. See *EI2*, *samā'*; and Karamustafa, 92.

6. See Crane, 1987; Sönmez, 1988.

7. Coomaraswamy, 1979, 15.

8. See Snodgrass, 1990, 1985; Critchlow, 1976; Bakhtiar, 1976; and Ardalan and Bakhtiar, 1973.

9. Guénon, 1975a, 73–74.

10. Coomaraswamy, 1979, 50–67.

11. Ibid., 64.

12. Ibid.

13. Schuon, 1984b.

14. Ibid., chap. 9.

15. Coomaraswamy, 1977, 2:6.

16. Ibid., 1:173, note 37; and 2:7.

17. Guénon, 1975b, xi.

18. Schuon, 1976a, vi. Nasr translates *philosophia perennis* as *al-ḥikma al-khālida* or *al-ḥikma al-laduniyya*.

19. Guénon, 1975b, x.

20. Schuon, 1975, 7.

21. Guénon, 1975a, 15–16

22. Ibid., 17.

23. Schuon, 1984a, 111.

24. Coomaraswamy, 1977, 1:13

25. Ardalan and Bakhtiar, 1973, xi.

26. Lings, 1991, viii.

27. *The Macquarie Dictionary*.

28. Coomaraswamy, 1980a, 15. On traditional symbolism, see also Coomaraswamy, 1980b, 1977; Eliade, 1991, 1986; and Snodgrass, 1990, 1:8–62.

29. Coomaraswamy, 1977, 1:324.

30. Ibid., 1980a, 15.

31. The following remarks on symbolism are based on Tillich, 1961a/b; Eliade, 1986, 1959a/b.

32. Snodgrass, 1985, 9.

33. Eliade, 1959b, 129. On the "nonhuman" origin of symbols, see Guénon, 1962.

34. Eliade, 1959a, 99; see also his, 1986, 5–6.

35. Ibid., 1986, 14.

36. Ibid., 1959a, 98.

37. Ibid., 1986, 6.

38. A. Coomaraswamy, 1956, 23.

39. Burckhardt, 1967b, 132; see also Nasr, 1987a, 3–14.

40. Burckhardt, 1967b, 133.

41. Necipoğlu, 1995, 80; see chap. 5.

42. Chorbachi, 1989.

43. Ibid., 757.

44. Ibid., 760.

45. Ibid.

46. Grabar, 1983, 27.

47. Ibid., 25.

48. Grabar, 1992, 236–37.

49. David Kolb, 1990, chap. 6.

50. Ardalan, 1980, illustrates the conventional understanding of symbolism my approach critiques.

51. Badawī, 1987. Primary and secondary sources on Sufism are numerous. For primary sources, see al-Hujwīrī, *Kashf al-Maḥjūb*; al-Sarrāj, *al-Luma*ʿ; al-Qushayrī, *Risāla*; al-Makkī, *Qūt al-Qulūb*. For secondary sources, see Knysh, 2000; Chittick, 2000; Nasr, 1979; Lings, 1975; Schuon, 1969, Schimmel, 1978.

52. Ibn Khaldūn, *Kitāb al-ʿIbar*, 2:863–82.

53. Al-Sarrāj, *al-Luma*ʿ, 20–21.

54. Al-Qushayrī, *Risāla*, 126–28.

55. Al-Sarrāj, *al-Luma*ʿ, 132–33.

56. Lings, 1975, 46.

57. Fritz Meier, 1976, 117–40.

58. Ibid., 118.

59. See Knysh, 2000, chap. 6.

60. See ibid., chaps. 7 and 8.

61. Ibid., 326.

62. Nasr distinguishes three types of writing on Sufism. He regards the works of the perennialists to be "of the greatest importance for an authentic understanding of Sufism," as their contribution represents "the truly authentic expositions of Sufism emanating from genuine teachings." 1980, 5.

63. A "holy tradition" is a prophetic tradition in which the speaker is God.

64. Al-Nābulusī, a follower of Ibn ʿArabī and a prolific scholar, is the author of numerous profound commentaries on eminent Sufi works, such as *Fuṣūṣ al-Ḥikam* of Ibn ʿArabī, the *Khamriyya* of Ibn al-Fariḍ, and *al-Nādirāt al-ʿAyniyya* of al-Jīlī.

65. Bukhārī, translation is based on Lings, 1975, 74.

66. Al-Nābulusī, "Munājāt" (MS. 5570), f.39.

67. For a full biography of Ibn ʿArabī see Addas, 1993. For an introduction to his life, works, and thought, see Chittick, 1989; Austin, 1980; Corbin, 1969; Nasr, 1964.

68. Ibn ʿArabī, *Fut.*, 1:153–54; see Chittick, 1989, xiii–xiv; Austin, 1980, 2–3.

69. ʿAfīfī, *Fus.*, 1:5; Yaḥyā, 2001, examines 994 titles attributed to Ibn ʿArabī.

70. *Fawāt al-Wafiyyāt,* cited in ʿAfīfī, *Fus.*, 1:6.

71. With each chapter including ten subchapters. Mentioned by Ibn ʿArabī in *al-Tadbīrāt*, 112, See Yaḥyā, 2001, 589.

72. Chittick, 1989, x.

73. Austin, 1980, 14.

74. Yaḥyā, 2001, 20.

75. Knysh, 1999.

76. Ibid., 1.

77. See for example, Chittick, 1998, 1994, 1989; Chodkiewicz, 1993; Corbin, 1969; Izutsu, 1983, 1971.

78. The *Futūḥāt* is now available in several Arabic editions. My references are to Dār Ṣādir's edition (4 vols.); however, I have consulted ʿUthmān Yaḥyā's, other editions and the manuscript copies of al-Ẓāhiriyya library for the cosmograms and critical passages.

79. MS. 123, 1:f.3/7.

80. See al-Suyūṭī, 1982, 9–14.

81. See al-Azraqī, 1983, 35.

82. *Ras.*, 2:276–77. See Nasr, 1978a, 45.

83. *Lis.*, "a.y.a."

84. Ibid., "r.m.z.," "sh.w.r."

85. Ibid., 1:173.

86. Ibid., 1:173. Chittick draws attention to the word *dalīl*, "evidence," "signifier," "reference," frequently used by Ibn ʿArabī in the sense of "proof" and discusses its conceptual relationship to "sign." 1998, 3–12.

87. *Fut.*, 1:174.

88. Snodgrass, 1985, 3.

89. *Fut.*, 1:551.

90. *Lis.*, "ʿ.b.r." For further treatment of the concept of 'crossing,' see *Fut.*, 1:207.

91. See al-Fārābī, *Kitāb al-Jamʿ*, Ibn Sīnā, *Shifāʾ*, and particularly the Shīrāzī's analysis in *al-Ḥikma al-Mutaʿāliya*, 2: chap. 9. On the Platonic doctrine of forms as the foundation of symbolism in the Western tradition, see Coomaraswamy, 1956; Snodgrass, 1990, 1:8–20.

92. *Lis.*, "m.th.l."

93. On *tanzīh* and *tashbīh*, see Chittick, 1998, xxi–xxii

94. *Fut.*, 3:314. See Chittick, 1998, 6–12.

95. *Fut.*, 1:206.

96. *Al-ishāra nidāʾ ʿalā raʾs al-buʿd wa bawḥ biʿayn al-ʿilla.* Ibn al-ʿArīf, *Maḥāsin*, 23. For an interpretation of this statement, see *Fut.*, 2:504; Abū Zayd, 1983, 266–68.

97. *Fut.*, 2:504.

98. See Guénon, 1984, 21.

99. See *Fut.*, 1:189; Abū Zayd, 1983, 267.

100. *Fut.*, 3:47.

101. Al-Qāshānī, *Sharḥ*, 113–14; see Izutsu, 1983, 90.

102. *Fut.*, 3:46. Ibn ʿArabī's concept of 'immutable essences' has been discussed in a number of studies. See Chittick, 1989, 83–88, Izutsu, 1983, 159–96; Abū Zayd, 1983, 75–86; Burckhardt, 1976a, 62–64; Afīfī, 1969.

103. See Izutsu, 1983, 159.

104. *Fus.*, 1:76; *Fut.*, 3:47.

105. *Fut.*, 3:518; 1:119.

106. Ibn ʿArabī, *Inshāʾ al-Dawāʾir*, 19. See Izutsu, 1983, 162.

107. *Fut.*, 3:47/225. See Chittick, 1989, 204.

108. *Fut.*, 3:47.

109. *Fus.*, 1:101. For more details, see *Fus.*, 2:110; Izutsu, 1983, 89; Corbin, 1969, 191.

110. *Ras.*, 2:6.

111. Ibid.

112. See the seventh and eighth tracts of the *Rasāʾil*.

113. *Ras.*, 1:277. On the premodern Islamic notion of art, see Burckhardt, 1967b; Nasr, 1981, chap. 8.

114. *Ras.*, 2:7–8.

115. Ibid., 2:6; see Nasr, 1978a, 58.

116. *Ras.*, 2:6–7; see Nasr, 1978a, 58–59.

117. *Ras.*, 1:277.

118. Nasr, 1978a, 59.

119. *Ras.*, 2:6–7; see Nasr, 1978a, 58–59.

120. *Ras.*, 2:7; see Nasr, 1978a, 59.

121. *Ras.*, 2:25; 3:235.

122. The Ikhwān interpret the Quranic verse: "Depart unto the shadow falling threefold" (77:30), as referring to the three-dimensional world of bodies. *Ras.*, 2:21.

123. *Ras.*, 2:53.

124. Ibid., 3:235–36.

125. The following discussion is based on my article, 1997b. For a thorough study of Ibn ʿArabī's concept of imagination, see Chittick, 1989, 1994; Corbin, 1972, 1969. Al-Nābulusī's teachings in the eighteenth century reveal the continuity and wide influence of Ibn ʿArabī's concept of imagination.

126. *Fut.*, 2:311/375. See Chittick, 1989, 115–21.

127. *Fut.*, 2:313.

128. *Ras.*, 3:416; *Fut.*, 2:691.

129. *Fut.*, 1:163/125–26.

130. *Miʿyār al-ʿIlm*, 90–92.

131. *Fut.*, 3: 46–47/518. On *barzakh*, see Chittick, 1989, 14–16, 117–18, 125–26; Burckhardt, 1979; Abū Zayd, 1983, 51–57.

132. *Fut.*, 3:46–47; 1:304; see Chittick, 1989, 117–18. Abū Zayd, 1983, 51–52.

133. *Fut.*, 2:129, 3:42; see Chittick, 1989, 15–16.

134. *Fut.*, 1:304.

135. *Fus.*, 104; see Izutsu, 1983, 7.

136. *Fut.*, 2:309–13; see Corbin, 1969, 219–20.

137. Corbin, 1969, 219.

138. Ibid., 219. Chittick refers to *al-khayāl al-muṭlaq* as "Nondelimited Imagination," to *al-khayāl al-munfaṣil* as "discontiguous imagination," and to *al-khayāl al-muttaṣil* as "contiguous imagination." 1989, 117.

139. *Fut.*, 2:311.

140. Ibid.; see Corbin, 1969, 219–20.

141. *Ras.*, 3:416. For a discussion of the notions of "infinity" and "indefinite" in relation to "possibility," see Guénon, 1984, 27–34. On art and imagination in premodern Islam, see Nasr, 1987a, 177–84.

142. *Fut.*, 2:309. The four degrees of existence are (1) *al-wujūd al-ʿaynī*, "the existence of a thing itself"; (2) *al-wujūd al-dhihnī*, "the existence of the thing in the mind"; (3) *al-wujūd al-lafẓī*, "the existence of the thing in utterance"; and (4) *al-wujūd al-raqmī*, "the existence of the thing in writing."

143. *Fut.*, 4:203.

144. Ibid.

145. *Ras.*, 1:277; 2:9.

146. Ibid., 1:399.

147. Ibid., 3:416.

148. Al-Ghazālī, *al-Risāla al-Laduniyya*, 21–22; see Smith, 1938, 2:191.

149. *Fut.*, 1:42.

150. Ibid., 1:42–43.

151. Ibid., 1:91.

152. *Lis.*, "ṣ.w.r."

153. The English word *art* derives from the Latin root *ars*, "to make." For more details, see Nasr, 1987a, 9; Coomaraswamy, 1956, 7–22; Michon, 1982, 49.

154. *Fut.*, 1:139.

155. Ibid., 1:140.

156. Al-Ghazālī, *al-Risāla al-Laduniyya*, 22; see Smith, 1938, 2:191.

157. Ibid., 22–24; see Smith, 1938, 2:191–92.

158. *Ras.*, 1:399.

159. These are *al-kawn wa al-fasād*, "generation and corruption," *al-ziyāda wa al-nuqṣān*, "increase and decrease," and *al-taghyīr wa al-naqla*, "change and translocation." *Ras.*, 2:25.

160. *Ras.*, 2:25.

161. Ibid.

162. This is similar to St. Thomas Aquinas' statement that "art imitates nature in her manner of operation," interpreted by Coomaraswamy as "God in His manner of creation," 1956, 34.

163. *Ras.*, 1:290.

164. Al-Ghazālī, *al-Maqṣad al-Asnā*, 162; introduction by Shehadi, xxvi.

165. Ibid., 182.

166. Ibid., introduction by Shehadi, l–li.

167. *Ras.*, 1:290.

168. Ibid.

169. Ibid.

170. *Fut.*, 1:119.

171. The Arabic term *taqdīr*, translated here as "design," comes from the verbal root *qaddara*, "to determine" and "to measure." The trilateral root "q.d.r." has two literal meanings: "power" and "fate." Identifying the act of creation with designing through the word *taqdīr* brings together at once the ideas of "determination" *(taqdīr)*, "measure" *(qadr)*, "power" *(qudra)*, and "predestination" *(qadar)*. See *Fut.*, 2:62

172. *Al-bāri*ʾ also means the "Animator," hence the association of the act of producing with that of giving life.

173. *Aḥsan* literally means "best" and "most beautiful," suggesting a close relationship in the context of making.

174. Al-Ghazālī, *al-Maqṣad al-Asnā*, 80.

175. Ibid..

176. *Fut.*, 4:212. Ibn ʿArabī considers plants and minerals to be living creatures, whose living nature is inconspicuous. *Fut.*, 3:437.

177. Al-Ghazālī, *al-Maqṣad al-Asnā*, 82.

178. Ibid., 83.

2. Metaphysical Order

1. Al-Ghazālī, *al-Maqṣad al-Asnā*, 82

2. For a critique of the Cartesian philosophy, see Heidegger, 1962, 122–34.

3. See Wu, 1998, 135–45.

4. *Ras.*, 1:78–79.

5. Ibid., 1:101.

6. Ibid., 1:79–80.

7. Ibid., 1:80.

8. Ibid.

9. On the Sufi concept of 'Unity of Being,' see al-Nābulusī, *al-Wujūd al-Ḥaqq*; Chittick, 1989, chap. 3; Izutsu, 1971, 1–24; Lings, 1971, 121–30, 1975, 63–73; Burckhardt, 1976a, 53–57.

10. In *al-Mabāḥith al-Mashriqiyya* al-Rāzī argues against this view.

11. On the concept of 'being' in Arabic, see Shehadi, 1975.

12. The Sufi concept of 'ecstasy,' *wajd*, also derives from the same root.

13. *Tuḥfa*, 47. See al-Nābulusī's commentary, *al-Qawl al-Matīn*.

14. See particularly the work of Ibn Taymiyya, 2000, 2:72–256

15. Lings, 1971, 128.

16. Al-Ghazālī, *Rasā'il*, 4:17; translation is based on Lings, 1971, 123.

17. *Al-Insān al-Kāmil*, 1:13. See Burckhardt, 1983, 3.

18. *Fut.*, 3:420.

19. For glossaries of Sufi terms, see Burckhardt, 1983, 59–73, 1976a, 114–26, 1975, 135–46.

20. *Tuḥfa*, 47. See al-Qāshānī, *Sharḥ*, 3.

21. See al-Jīlī, *al-Insān al-Kāmil*, 1:13–16. On the nature of this state, see Burckhardt, 1983, 4, 1976a, 114; Schuon, 1969, 142/144; Guénon, 1984, 43–49.

22. *Tuḥfa*, 48. On the idea of "Muḥammadan Reality," see Ibn ʿArabī, *Majmūʿat Rasā'il*, 1:374–454; *Fut.*, 1:119/143–44; *Fus.*, 1:214–26; Abū Zayd, 1983, 86–95.

23. *Tuḥfa,* 48.

24. Ibid.

25. Ibid.

26. Ibid.

27. Ibid.

28. Ibid. For a commentary on the seven states of Being, see al-Nābulusī, *al-Qawl al-Matīn*, 11–17. Schuon compares the seven states of universal manifestation to the "five divine presences"—*nāsūt*, *malakūt*, *jabarūt*, *lāhūt*, and *hāhūt*—in which each of *nāsūt* and *lāhūt* corresponds to two states. 1969, 142–58.

29. On the Sufi concept of 'divine business,' see Ibn ʿArabī, "Ayyām al-Sha'n," in *Majmūʿat Rasā'il*, 1:503–19; al-Qāshānī, *Sharḥ*, 4.

30. *Tuḥfa,* 49–50.

31. Al-Ḥallāj, *Ṭawāsīn*, 71.

32. Ibid., 51.

33. See *al-Isfār al-Gharīb.*

34. *Ras.*, 1:80.

35. *Kahf*, 3.

36. *Fut.*, 1:41.

37. Lings, 1971, 149.

38. *Kahf*, 3.

39. On negation as a mode of knowledge, see *Fut.*, 1:91–93.

40. Ibn ʿArabī, 1976, 3.

41. See Coomaraswamy, 1977, 2:220–30; Guénon, 1984, 48.

42. *Tuḥfa*, 47–52.

43. Ibid., 50. See Izutsu, 1971, 45; Bakhtiar, 1976, 10.

44. *Fut.*, 3:275–76.

45. Ibn ʿArabī says that the second part of this *ḥadīth* is a later addition. *Fut.*, 2:56.

46. See Ibn ʿArabī, 1976, 3.

47. See *Fut.*, 3:290.

48. Ibid., 1:260.

49. Ibid., 1:259.

50. Ibid., also 1:46.

51. Ibid., 3:275.

52. See Izutsu, 1983, 199.

53. *Fus.*, 1:115; see Izutsu, 1983, 198–99.

54. *Fut.*, 1:260.

55. Ibid., 4:14.

56. Ibid., 3:275.

57. Ibid.

58. See ibid., 4:193.

59. Ibid., 3:275.

60. Ibid., 1:255.

61. Ibid., 3:420.

62. Ibid., 1:293.

63. See Izutsu, 1983, 99.

64. *Fut.*, 3:197–98.

65. Al-Qāshānī, *Bayān*, 90.

66. *Fut.*, 1:293.

67. Ibid.

68. Ibid.

69. Ibid., 1:430.

70. This analogy is based on a prophetic tradition. See *Fut.*, 1:293.

71. *Fut.*, 1:293.

72. Ibid.

73. Ibid.

74. *Fus.*, 1:220. Ibn ʿArabī is referring here to the Arabic language. It is interesting to note that the Arabic term *nuqṭa*, "point," is also of feminine gender. On his metaphysical interpretation of gender, see *Fut.*, 1:138–43; Burckhardt, 1975, 124.

75. *Fut.*, 1:259.

76. Ibid., 1:260.

77. Ibid., 1:388.

78. Ibid., 4:210.

79. Ibn ʿArabī, *ʿUqlat al-Mustawfiz*, 45; *Fut.*, 1:90.

80. *Fus.*, 1:115; see Izutsu, 1983, 198–200.

81. *Tuḥfa,* 48–49.

82. *Fut.*, 1:100.

83. That is, two premises and a conclusion. See Akkach, 1997b, 101.

84. *Fut.*, 3:276.

85. Ibid., 2:493.

86. On the life of al-Ḥallāj, see Massignon, 1982; Massignon and Kraus, 1936; Makārim, 1989.

87. *Fus.*, 1:83.

88. Ibid., 1:49; see Izutsu, 1983, 227.

89. For detail, see *Sharḥ*, 266.

90. See *Fus.*, 1:214; Burckhardt, 1975, 116–17.

91. *Sharḥ*, 266; see Izutsu, 1983, 237.

92. *Fus.*, 48–49; see Burckhardt, 1975, 8–9. On this, see Nasr, 1968c, 1980, 31–39, 1981, 160–88; Izutsu, 1983, 243–83; Burckhardt, 1983.

93. See *Fut.*, 3:190.

94. *Sharḥ*, 11; see Izutsu, 1983, 222.

95. *Tuḥfa,* 48.

96. Burckhardt translates selected excerpts, 1983.

97. *Al-Insān al-Kāmil*, 1:5.

98. Ibid., 2:46.

99. *Inshāʾ al-Dawāʾir*, 21. See *Fus.*, 1:50; Izutsu, 1983, 235.

100. *Inshāʾ al-Dawāʾir*, 22.

101. *Fut.*, 2:446.

102. See Guénon, 1975b; Snodgrass, 1985, 21.

103. *Fut.*, 1:124/153.

104. Ibid., 3:198.

105. Ibid.

106. Ibid., 1:123.

107. Ibid., 2:445.

108. Ibid., 2:464. See Ibn Kathīr, *Tafsīr*, 3:240.

109. See Ibn Kathīr, *Tafsīr*, 4:295.

110. *Kitāb al-ʿAyn*, 1:235; see also *Lis.*, "ʿ.j.b."

111. The notion of the sacred bone was present in many ancient and premodern traditions, see Stross, 1994.

112. *Lis.*, "n.sh.ʾ"

113. *Fut.*, 1:312.

114. Ibid., 2:464.

115. *Shifāʾ*, 8:400.

116. *Ras.*, 1:183–241.

117. Ibid., 1:223–24. This is strikingly similar to Vitruvius' proportional system, 1960, 72–75.

118. *Fus.*, 1:203–04; see Burckhardt, 1975, 104–05.

119. *Sharḥ*, 279.

120. Movements caused by external forces such as gravity, for example, are not included here.

121. *Fut.*, 2:464–65.

122. Ibid., 4:315.

123. On the philosophical treatment of the human directions, see Ibn Sīnā, *Shifāʾ—Ṭabīʿiyyāt*, 1:246–58.

124. See Ṭabarī, *Jāmiʿ*, 5:345–47.

125. *Fut.*, 1:53.

126. Al-Tusturī, *Tafsīr*, cited in Böwering, 1980, 149.

127. Böwering, 1980, 153. On this, see also al-Makkī, *ʿIlm al-Qulūb*, 94; Daylamī, *ʿAṭf al-Alif*, 33.

128. Böwering, 1980, 152; al-Tusturī, *Tafsīr*, 41; al-Makkī, *ʿIlm al-Qulūb*, 93–94; Daylamī, *ʿAṭf al-Alif*, 33.

129. Böwering, 1980, 154.

130. Daylamī, *ʿAṭf al-Alif*, 33; Böwering, 1980, 150.

131. Daylamī, *ʿAṭf al-Alif*, 33; Böwering, 1980, 150.

132. For further treatment of this idea, see *Fut.*, 1:302.

133. *Fut.*, 1:158.

134. Ibid., 4:76; 1:160.

135. Ibn ʿArabī, *Tafsīr*, 1:426.

136. Schuon, 1976b, 60. See also Nasr, 1979, 41; and Massignon, 1982, 2:3.

137. See *Fut.*, 1:51–91.

138. Ibid., 1:101.

139. See Lings, 1971.

140. Al-Qūnawī, *Mirʾāt al-ʿĀrifīn*, 24. The Ẓāhiriyya MS. 4865, 24–29, attributes the treatise to Ibn ʿArabī. On the world as a book, see *Fut.*, 2:395–478; 4:166–67.

141. Ibn ʿArabī calls this science "the science of Christhood" *(al-ʿilm al-ʿīsawī)*, referring to the Quranic description of Christ as "his word which he conveyed unto Mary, and a spirit from him" (4:171). See *Fut.*, 1:167–69.

142. This is according to a Sufi tradition that says: "God created the universe through the Breath of the Compassionate." See *Fut.*, 2:395–403. Burckhardt, 1976a, 43; Izutsu, 1983, 132.

143. This refers to the disjointed letters of the Quran, standing in isolation yet in a textual context.

144. *Mirʾāt al-ʿĀrifīn*, 25; see Askari's translation, 21.

145. *Mirʾāt al-ʿĀrifīn*, 24–25; see Askari's translation, 22; and see Schuon's commentary, 1976b, 64.

146. *Kahf,* 2; see Lings, 1971, 148.

147. For further detail, see *Fut.*, 1:111, 2:134–39.

148. *Mir'āt al-ʿĀrifīn*, 8–12.

149. *Jaw.*, 10/6.

150. Ibid., 22/12.

151. *Kahf*, 5. See "Kitāb al-Alif" in *Majmūʿat Rasā'il*, 1:491–502.

152. Lings, 1971, 154.

153. *Jaw.*, 14.

154. Lings, 1971, 156.

155. *Kahf*, 3.

156. Lings, 1971, 153–54. See *Fut.*, 2:123.

157. *Kahf*, 3.

158. Lings, 1971, 153–54.

159. Ibid., 148. See *Kahf*, 2. The same *ḥadīth* is quoted by al-Ḥallāj; see Massignon and Kraus, 1936, 95. See also "Kitāb al-Bā'," in *Majmūʿat Rasā'il*, 1:455–67.

160. Lings, 1971, 155.

161. Ibid., 155.

162. Ibid., 150.

163. Ibid., 157.

164. Ibid., 152.

165. Ibid., 151.

166. *Fut.*, 1:102.

167. *Kahf*, 4.

168. *Fut.*, 1:84.

169. *Fus.*, 1:49; see Burckhardt, 1975, 9.

170. Grammatically, the letter *alif* (A) does not admit any of the "phonetic motions" *(ḥarakāt)* and as such is unpronounceable. At the beginning of a word it is always detached from subsequent letters. These form a part of its symbolism; see *Fut.*, 1:61.

171. *Fut.*, 1:62.

172. Ibn al-Nadīm, *al-Fihrist*, 15. These movements correspond to Ibn ʿArabī's movements of spatial unfolding.

173. *Fut.*, 1:84–85.

174. Lings, 1971, 153.

175. See Guénon, 1975b, 46. On the symbolism of the tree, see Reat, 1975; Cook, 1988.

176. *Fut.*, 2:379; See Corbin, 1969, 188.

177. *Kun*, an imperative of *kāna*, should be *kuwn*, but the letter *w* is grammatically omitted because it lies between two consonants. *Kuwn* (be) and *kawn* (cosmos) have the same three letters *k.w.n.* See, Jeffrey, 1959.

178. *Shajarat*, 39–55; translation is based on Jeffrey, 1959, 62–72.

3. Cosmic Order

1. *ʿUqlat al-Mustawfiz*, 51–52. For more detail, see *Fut.*, 3:296, 2:67.

2. *Fut.*, 3:363.

3. Cited in Ardalan and Bakhtiar, 1973, 2.

4. *Fut.*, 3:416–48.

5. Corbin refers to this as "the Sigh of existentiating Compassion," 1969, 185.

6. See Izutsu, 1983, chap. 6, 152–58.

7. Ibid., 155. For more detail, see *Sharḥ*, 10.

8. See *Fut.*, 3:420, 2:310. Insofar as the divine Breath signifies the primordial substance, the *materia prima* of the world, it is the root of "otherness." See Burckhardt, 1986, 63.

9. Izutsu, 1983, 131–32.

10. Ibid., 132.

11. *Ḥadīth qudsī*, trans. Corbin, 1969, 184.

12. *Sharḥ*, 182.

13. *Fus.*, 1:112; see Izutsu, 1983, 132; Corbin, 1969, 184.

14. *Sharḥ*, 182; *Fut.*, 2:331.

15. *Fut.*, 2:434.

16. *Fus.*, 1:145; see Burckhardt, 1975, 77.

17. *Fut.*, 2:331; 3:452.

18. *Fus.*, 1:144.

19. *Fut.*, 2:310; see Corbin, 1969, 185.

20. *Fut.*, 3:430.

21. Ibid., 2:310.

22. Corbin, 1969, 185.

23. *Fut.*, 1:119.

24. Ibid., 1:305.

25. Ibid., 1:41.

26. Ibid., 2:311.

27. Burckhardt compares Ibn ʿArabī's immutable essences to the Platonic ideas, 1976a, 62–64.

28. *Fut.*, 4:315.

29. Ibid., 2:421; see also 1:90–91.

30. *Fut.*, 3:420.

31. Ibid., 3:416–48.

32. See Abū Zayd, 1983, 97–149.

33. The Soul's practical faculty is responsible for the manifestation of forms, whereas the intellectual faculty is responsible for knowing their measures and proportions. See *Fut.*, 2:429.

34. Also referred to as *al-habāʾ* and as *al-ḥaqīqa al-kulliyya* (Universal Reality), see *Fut.*, 1:118.

35. Murata, 1987, 330.

36. *Fut.*, 1:148.

37. Ibid., 1:199.

38. Ibid., 3:420.

39. See *Fus.*, 1:219; Izutsu, 1983, 134.

40. *Fut.*, 3:420.

41. *Fut.*, 3:420, see also *Fus.*, 1:144.

42. Al-Jurjānī, *Taʿrīfāt*, 319.

43. "*Substansia*, from *sub stare*, is literally 'that which stands beneath', a meaning also attached to the ideas of 'support' and 'substratum'." Guénon, 1983b, 26.

44. See *Ras.*, 2: tract 1; Nasr, 1978a, 58–61.

45. *Fut.*, 2:439.

46. Ibid., 1:55.

47. See Schuon, 1969, 103–04.

48. Ibid., 102. There are many versions of this *ḥadīth*; see al-Ṭabarī, *Tārīkh*, 1:16–19; Ibn Kathīr, *al-Bidāya*, 1:8–9; al-Suyūṭī, *al-Hay'a*, 6–9.

49. Schuon, 1969, 103–04.

50. *Fut.*, 2:431.

51. Ibid., 1:260.

52. *'Uqlat al-Mustawfiz*, 56–57.

53. See Jurjānī's, *Ta'rīfāt*, 295.

54. See *Fut.*, 1:147.

55. Ibid., 2:436.

56. Ibid., 3:431.

57. *Al-Bidāya*, 1:11.

58. *Fut.*, 2:436, 1:44.

59. Ibid., 3:432.

60. See al-Suyūṭī, *al-Hay'a*, 3.

61. With reference to Ibn Masarra, Ibn 'Arabī affirms these forms of the bearers, which correspond to those in Ezekiel's vision. See *Fut.*, 1:149.

62. Schuon, 1969, 109.

63. Mālik is known in medieval Islamic literature as an angel and not a prophet.

64. *Fut.*, 3:432.

65. Ibid., 1:147–48.

66. Ibid., 3:432.

67. Al-Suyūṭī, *al-Hay'a*, 1–5.

68. *Fut.*, 3:432.

69. Schuon, 1969, 111–20.

70. See *Fut.*, 2:436.

71. Schuon, 1969, 119.

72. *Fut.*, 3:433.

73. "Tashrīḥ al-Aflāk," 2.

74. *Fut.*, 3:435.

75. Ibid., 2:441; see also 1:317–18.

76. Ibid., 2:550.

77. Ibid., 2:441.

78. Ibid.

79. Ibid., 3:433.

80. *Al-Hay'a*, 10.

81. Ibid., 10

82. *Fut.*, 3:437.

83. See Guénon, "The Mysteries of the Letter *Nūn*," 1947, 166–68.

84. Akkach, 1997b, 101.

85. *Fut.*, 1:107.

86. *Al-Hay'a*, 34–35.

87. Ibid., 34.

88. *ʿUqlat al-Mustawfiz*, 7.

89. See *Fut.*, 2:677.

90. Cited in Wensinck, 1978, 20.

91. Ibid., 20.

92. Ibid.

93. *Tafsīr*, 3:93.

94. Corbin, 1969, 56. See also Coomaraswamy, 1970.

95. Al-Āmidī, *Ris.*, 9–10.

96. See Ibn Kathīr, *al-Bidāya*, 1:92; Ibn Athīr, *al-Kāmil*, 1:37. Ibn Kathīr favors a version of this *ḥadīth* that says Adam was sixty cubits in height, and his progeny was progressively shortened until the time of the Prophet. 1:85–92.

97. According to *Lis.*, the Arabic term *ʿamad* refers to "the columns upon which a house stands," and to "the perpendicular timber pole at the center of a nomadic tent *(khibā')*."

98. See *Fut.*, 1:138–43.

99. Ibid., 4:416.

100. Al-Āmidī, *Ris.*, 16.

101. See *Fut.*, 1:140.

102. *ʿUqlat al-Mustawfiz*, 56–57.

103. *Fut.*, 2:433.

104. *Fut.*, 3:548.

105. *Ras.*, 2:15; see Nasr, 1978a, 65.

106. *Ras.*, 2:16; see also, *Fut.*, 2:434.

107. *Fut.*, 2:434.

108. Ibid., 2:433.

109. According to Ibn ʿArabī, the day determined with reference to the divine Feet starts with Gemini, *Fut.*, 2:437–38; the day determined with reference to the sphere of the fixed stars starts with Libra, *ʿUqlat al-Mustawfiz*, 61; and the day determined with reference to the sun starts with Aries, *Fut.*, 3:461, 2:438.

110. *Fut.*, 1:121.

111. Ibid., 2:443.

112. Ibid., 2:437–38.

113. The names of the first five days of the week derive from the Arabic names of the first five numerals, 1 to 5, and correspond to the five stars of retrograde motion. *Fut.*, 1:643.

114. Al-Qāshānī, *Bayān*, 91.

115. See Nasr, 1981, 202; Burckhardt, 1977, 44.

116. See *Fut.*, 1:122/291.

117. Al-Qāshānī, *Bayān*, 90–91.

118. Burckhardt, 1977, 35. On the cosmological symbolism of the Arabic alphabet, see *Fut.*, 1:51–91.

119. Burckhardt, 1977, 35.

120. Ibid., 35.

4. Architectural Order

1. Al-Ghazālī, *Majmūʿat Rasāʾil*, 1:5–6.
2. Crane, 1987, 19–20.
3. Papadopoulo, 1980, 244–45.
4. On medieval tomb towers, see Daneshvari, 1986.
5. See Lehrman, 1980.
6. Ardalan and Bakhtiar, 1979, 68.
7. Ibid., 89.
8. Burckhardt, 1976b, 179.

9. *Ras.*, 1:78. See Nasr, 1978a, 40–43.

10. Geometry in premodern Islamic art and architecture is a vast field, different aspects of which have been covered by many studies. See, for example, Necipoğlu, 1995; Walls, 1990; Paccard, 1980; Critchlow, 1976; El-Said and Parman, 1976.

11. *Ras.*, 1:102. See Critchlow, 1969, 4–5; 1976, 9–14.

12. Ibid. 1:103–04.

13. On Eliade's concept of 'sacred space,' see his, 1986, 107–24, 1974, 1–37/367–87, 1959b, 20–65.

14. Melissa Raphael, 1997, 1.

15. Rudulf Otto, 1926, vii.

16. Ibid., 26.

17. Ibid.

18. Ibid., 28. See Akkach, 2003.

19. Eliade, 1959b, 10.

20. Ibid., 11.

21. Ibid., 20–113.

22. See Eliade, 1994.

23. Otto, 1926, 5.

24. Eliade, 1959b, 20.

25. See al-Ṭabarī, *Jāmiʿ al-Bayān,* 8:396–97.

26. See Akkach, 2002b.

27. See *EI2*, *Faḍīla*.

28. Ibn Manẓūr, *Mukhtaṣar*, 1:255.

29. Ibid., 1:255–59.

30. See al-Wāsitī, *Faḍāʾil*. Dating from the early eleventh century, this is among the earliest surviving texts on the merits of Jerusalem. For a detailed study of this genre, see Ibrāhīm, 1985.

31. Eliade, 1974, 369.

32. Dickie, 1978, 33.

33. Ibn al-Athīr, *al-Kāmil*, 2:527–30; Ibn Kathīr, *al-Bidāya*, 7:74–75; Al-Ṭabarī, *Tārīkh*, 2:188–94. In describing the layout, Ibn al-Athīr refers to both cities whereas the other two sources refer to al-Kūfa only.

34. Al-Ṭabarī, *Tārīkh*, 2:194–95; al-Baghdādī, *Tārīkh*, 1: 66–79; Ibn al-Athīr, *al-Kāmil*, 2:530.

35. Al-Baghdādī, *Tārīkh*, 1:72.

36. Al-Maqrīzī, *Khiṭaṭ,* 4:36.

37. Al-Baghdādī, *Tārīkh*, 1:67. According to Ardalan and Bakhtiar, Kāshān was founded under the ascendency of Virgo, 1973, 136, note 8.

38. A-Maqrīzī, *Khiṭaṭ*, 2:204.

39. The term *qibla* derives from *al-qubl*, "man's front," see *Lis.*, "q.b.l."

40. In several publications David King has uncovered the rich Islamic tradition associated with orientation. See King, 1982 and 1999.

41. Ibn ʿArabī, "al-Aḥadiyya" (MS. 123, Zah.), 18.

42. Snodgrass, 1985, 32.

43. Ibid., 30–36. See Eliade, 1959b, 29.

44. From *sharaqa*, "to shine," "to illumine," "to radiate."

45. Presuming that the spectator is facing south.

46. *Fut.*, 1:387. West, *maghrib*, from *gharaba*, "to go away."

47. Al-Qāshānī, *Bayān*. See Akkach, 1997a.

48. For a philosophical discussion of this, see Ibn Sīnā, *Shifāʾ*: *al-Samāʿ al-Ṭabīʿī*, 251–58.

49. See King, 1982, 303–12.

50. On the names and meanings of the directions, see *Lis.*

51. Al-Balādhurī, *Futūḥ*, 274–75; Al-Ṭabarī, *Tārīkh*, 1:191.

52. The first ritual prayer starts before sunrise and the last after the sunset.

53. *Fut.*, 1:397.

54. See, for example, Ikhwān, *Ras.*; al-Muqaddasī, *Aḥsan al-Taqāsīm*; and al-Maqrīzī, *Khiṭaṭ*.

55. Corbin, 1980b, 138.

56. Ibid., 141.

57. See, for example, Ibn Qayyim al-Jawziyya, *Ḥādī al-Arwāḥ*.

58. Ibn ʿArabī, 1986, 244.

59. *Fut.*, 1:126–31. For more detail, see Rahman, 1964, 167–80; Corbin, 1969, 351, and note 10.

60. *Fut.*, 127; translation is based on Corbin, 1969, 351, note 10.

61. *Fut.*, 1:130–31.

62. Corbin, 1969, 190.

63. *Fut.*, 1:126; see Rahman, 1964, 173.

64. *Fut.*, 1:127; translation by Corbin, 1969, 351.

65. *Fut.*, 1:127. On Ibn ʿAbbās, see Ibn Saʿd, *Ṭabaqāt*, 2:365–72.

66. Rahman, 1964, 173.

67. *Fut.*, 1:128; Rahman, 1964, 173.

68. *Fut.*, 1:129.

69. See Stewart, 1980, 33.

70. *Fut.*, 1:130.

71. In another tradition, God lifted Abraham up to heaven so that he may look down at the whole expanse of earth and choose the spot for the house. Once he did, the angels told him that he has chosen the sanctuary *(ḥaram)* of God on earth. See Al-Azraqī, *Akhbār,* 1:53.

72. There are many versions of this tradition, see Al-Azraqī, *Akhbār*, 1:58–66; al-Ṭabarī, *Tārīkh*, 1:128–34.

73. Al-Tirmidhī, *Khatm al-Awliyāʾ*, 350, 372.

74. *Fut.*, 2:60.

75. See Guénon, 1975, 45.

76. *Fut.*, 1:666.

77. Cited in Wensinck, 1978, 36.

78. Wensinck, 1978, includes English translations of some of these traditions.

79. See King, 1993, x; King and Lorch, 1992.

80. *Ras.*, 2:39.

81. See al-Azraqī, *Akhbār*, 1:32; Wensinck, 1978, 36.

82. Al-Azraqī, *Akhbār,* 1:31; Wensinck, 1978, 18.

83. In yet another tradition, the first layout of the house was done by the angels who were ordered by God to build for the creatures of the earth a house exactly in the likeness and measure of their heavenly one, *al-ḍurāḥ*. Al-Azraqī, *Akhbār*, 1:33–34/36–37.

84. Al-Yaʾqūbī, *Tārīkh*, 2:19. The symbolism of the Kaʿba as the navel of the earth has been discussed in Wensinck, 1978. On the symbolism of the omphalos, see also Eliade, 1974, 231–35.

85. Al-Bīrāmī, *Yad*, 104. On al-Bīrāmī, see al-Ziriklī, *al-Aʿlām*, 4:101.

86. Ibid., 104.

87. Cited in Wensinck, 1978, 47.

88. Ibid., 47–48.

89. Al-Azraqī, cited in Wensinck, 1978, 14–15.

90. Ibid., 14.

91. Al-Azraqī, *Akhbār*, 1:35; translation is based on Wensinck, 1978, 51–52.

92. The Frequented House is believed to have first been in the primordial paradise; it was brought down to earth for the sake of Adam and then at the time of the deluge was lifted up again to remain permanently in heaven till the time of the resurrection.

93. Al-Azraqī, *Akhbār*, 1:49; see Wensinck, 1978, 48.

94. *Fut.*, 2:443.

95. At the times of draught, the people of Arabia used to send messengers to Mecca to pray, who were usually advised to climb the mount of Abu Qubays. See Wensinck, 1978, 25. On the idea of the mountain as primordial center and the complementary symbolism of the mountain and the cave, see Guénon, 1971a/b.

96. Al-Azraqī, *Akhbār*, 1:31–32.

97. Ibid., 1:43.

98. Al-Ṭabarī, quoted in Wensinck, 1978, 42.

99. Al-Azraqī, *Akhbār*, 1:43; see also Wensinck, 1978, 50.

100. Al-Āmidī, *Ris.*, 17; Wensinck, 1978, 42.

101. Ibn al-Wardī, cited in Wensinck, 1978, 40.

102. Al-Qazwīnī, cited in Wensinck, 1978, 35.

103. *Fut.*, 2:6–7

104. Ibid., 2:6. The *shīʿī* symbolism of the Kaʿba incorporates similar meanings. See Corbin, 1965, 79–166.

105. *Fut.*, 2:6–7.

106. Ibid., 1:160.

107. See *Ras.*, 1:229–32.

108. *Fut.*, 3:198.

109. Ibid., 1:665; Al-Bīrāmī, *Yad*, 95.

110. From the Sufi perspective, the black stone also represents *al-mīthāq*, the "covenant" between God and man, whereby the Kaʿba is distinguished from its supernal and infernal models.

111. Al-Bīrāmī, *Yad*, 95. According to a Sufi interpretation, during the ritual circumambulation of the Kaʿba the Muslim is supported by two right hands: his own and

God's represented by the black stone that stands to the left-hand side. See Al-Bīrāmī, *Yad*, 102; *Fut.*, 1:749.

112. *Fut.*, 1:50.

113. Al-Bīrāmī, *Yad*, 101.

114. Al-Azraqī, *Akhbār*, 1:170–71; *Fut.*, 4:32; Ibn ʿArabī, *Muḥāḍarat al-Abrār*, 1:266.

115. *Fut.*, 1:666.

116. Ardalan and Bakhtiar, 1979, 29; Burckhardt 1976b, 4.

117. See *Lis.*, "k.ʿ.b."

118. *Fut.*, 1:666.

119. *Asās al-Balāgha*, 394.

120. Al-Farāhīdī, *Kitāb al-ʿAyn*, 1:207.

121. Ibn ʿArabī considers the undetermined corner of *al-ḥijir* to be "the real house of God." *Fut.*, 1:51/705.

122. *Fut.*, 1:703; al-Azraqī, *Akhbār*, 1:335–36.

123. *Fut.*, 1:50.

124. Ibid.

125. Ibid., 1:666.

126. Ibid., 1:666–67.

127. Ibid., 1:706.

128. Mosque architecture and its form, historical development, sociocultural contexts, and symbolism have been discussed in numerous studies. See, for example, Frishman and Khan, eds., 1994; Hillenbrand, 1994, chapter 2; al-Walī, 1988; Nasr, 1987a, 37–63; Dikie, 1987, 15–47; Burckhardt, 1976b, 18–25.

129. This is more conspicuous in the early periods.

130. Some narratives do exist though; see Ibn Saʿd, *Ṭabaqāt*.

131. Eliade, 1974, 369.

132. Lings, 1983, 123–24.

133. Ibn Kathīr, *Al-Bidāya*, 3:215; Ibn Saʿd, *Ṭabaqāt,* 1:240.

134. *Lis.*, "ʿ.r.sh."

135. Frishman, 1994, 32–33.

136. Ibn Saʿd, *Ṭabaqāt*, 1:239–41.

137. *Fus.*, 1:222–23; see Burckhardt, 1975, 127–28.

138. See Corbin, 1969, 249.

139. *Fus.*, 1:83; translation by Corbin, 1969, 254.

140. Ibn ʿArabī, "Mirʾāt" (MS. 4865, Zah.).

141. Corbin, 1969, 251. See *Fus.*, 2:341–42.

142. *Fut.*, cited in Burckhardt, 1967b, 134.

143. *Fut.*, 3:63.

144. Al-Ghazālī, *Iḥyāʾ*, 1:153.

145. *Fus.*, 1:223; Burckhardt, 1975, 128.

146. *Fus.*, 1:223.

147. The following remarks on acoustic space are based on Ong, 1967a, 160–69; and Carpenter and McLuhan, 1966, 65–70.

148. Ong, 1967a, 164.

149. Ibid., 163; Carpenter and McLuhan, 1966, 67.

150. Ong, 1967a, 164.

151. Ibid., 1967b, 7.

152. Carpenter and McLuhan, 1966, 67.

153. See al-Ghazālī, *Iḥyāʾ*, 1:164.

154. Burckhardt, 1976b, 5.

155. Ibid., 18–19; 1967b, 136.

156. Grabar, 1976, chapter 4, 35.

157. Burckhardt, 1976b, 19.

158. Ibid., 1967b, 136.

159. Ong, 1967a, 164.

160. *Fus.*, 1:224; see Burckhardt, 1975, 130.

161. *Fut.*, 2:486.

162. Ibid., 1:426.

163. Ibid., 1:704–05.

164. Ibid., 1:139.

165. Ibid., 2:366–68.

166. Ibid., 2:366.

167. Ibid., 2:367–68.

Afterword

1. Rees, 2001, 141.
2. Davies, 1992, 20–21.
3. Rees, 2001, 180.
4. See ibid., 200, 8.
5. Gleiser, 2001, 235.
6. Rees, 2001, 153.
7. Davies, 1992, 232.
8. Harries, 2001, 331.

List of Arabic Manuscripts Cited

The following listing is according to the indices of Dār al-Kutub al-Ẓāhiriyya.

Al-Āmidī, Ibrāhīm al-Qarmānī. MS. 6102. "Risāla fī Bayān Abʿād al-Samāwāt wa al-Ardīn wa al-ʿArsh wa al-Lawḥ wa al-Qalam wa ʿAjāʾib al-Samāwāt wa Makān al-Janna wa al-Nār," f. 1–32.

Al-ʿĀmilī, Bahāʾ al-Dīn. MS. 3103. "Tashrīḥ al-Aflāk," f. 1–11.

Al-Bīrāmī, ʿAbd Allāh ʿAbdī. MS. 7847. "Al-Yad al-Ajwad fī Istilām al-Ḥajar al-Aswad," f. 95–116.

Al-Hindī, Muḥammad b. Faḍl Allāh al-Barhānbūrī. MS. 5511. "Al-Tuḥfa al-Mursala fī Waḥdat al-Wujūd," f. 47–52.

Ibn ʿArabī, Muḥyī al-Dīn. MS. 123. "Inshāʾ al-Dawāʾir wa al-Jadāwil," f. 5–20.

———. MS. 1328. "Al-Futūḥāt al-Makkiyya," vol. 1.

———. MS. 1330. "Al-Futūḥāt al-Makkiyya," vol. 3.

———. MS. 3975. "Inshāʾ al-Dawāʾir wa al-Jadāwil," f. 4–15.

———. MS. 4865. "Mirʾāt al-ʿĀrifīn fi Multamas Zayn al-ʿĀbidīn," f. 24–29.

Ibn Sabʿīn, Abu Muḥammad Qutb al-Dīn (Attributed). MS. 7127. "Jawāhir al-Sirr al-Munīr," f. 1–262.

Al-Jīlī, ʿAbd al-Karīm. MS. 4992. "Al-Kahf wa al-Raqīm fī Sharḥ Bism Allāh al-Raḥmān al-Raḥīm," f. 1–11.

al-Nābulusī, ʿAbd al-Ghanī. MS. 5570. "Munājāt al-Hakīm wa Munāghāt al-Qadīm," f. 39–40.

Al-Qāshānī, ʿAbd al-Razzāq. MS. 5258. "Bayān Miqdār al-Sana al-Sarmadiyya wa Taʿyīn al-Ayyām al-Ilāhiyya," f. 90–93.

Al-Suyūṭī, Jalāl al-Dīn. MS. 5881. "Al-Hayʾa al-Saniyya fī al-Hayʾa al-Sunniyya," f. 9–26.

Selected Bibliography

Abū Zayd, Naṣr Ḥāmid. 1996. *Al-Ittijāh al-ʿAqlī fī al-Tafsīr*. Beirut: Al-Markaz al-Thaqāfī al-ʿArabī, 3rd ed.

———. 1983. *Falsafat al-Taʾwīl*. Beirut: Dār al-Tanwīr.

Ackerman, P. 1953. "The Symbolic Sources of Some Architectural Elements." *JSAH* 12, 4: 3–7.

Addas, Claude. 2000. *Ibn ʿArabī: The Voyage of No Return*, trans. D. Streight. Cambridge: The Islamic Texts Society.

———. 1993. *Quest for the Red Sulphur: The Life of Ibn ʿArabī*, trans. P. Kingsley Cambridge: The Islamic Texts Society, rep. 2000.

ʿAfīfī, A. A. 1969. "Al-Aʿyān al-Thābita fī Madhhab Ibn ʿArabī wa al-Maʿdūmāt fī Madhhab al-Muʿtazila." In Ibrahīm Madkūr, ed., 1969.

———. 1939. *The Mystical Philosophy of Muhid Dīn Ibnul-ʿArabī*. Cambridge: Cambridge University Press.

Akkach, Samer. 2003. "The 'Wholly Other': On the Spatiality of the Holy in Islam." *ATR* 8, 2: 41–58.

———. 2002a. "Religious Mapping and the Spatiality of Difference." *Thresholds* 25: 68–75.

———. 2002b. "Mapping Difference: On the Islamic Concept of *faḍāʾil*." In S. Akkach, ed., *De-Placing Difference: Architecture, Culture and Imaginative Geography*. Adelaide: CAMEA, 9–21.

———. 1997a. "Ibn ʿArabi's Cosmogony and the Sufi Concept of Space and Time." *Disputatio* 2: 115–42

———. 1997b. "The World of Imagination in Ibn ʿArabi's Ontology." *BJMES* 24: 1, 97–113.

ʿAlī, Abdullah Y. 1983. *The Holy Qur'an: Text, Translation and Commentary*. Lahore: Ashraf.

Al-Anṣārī, Shams al-Dīn. 1923. *Nukhbat al-Dahr fī ʿAjāʾib al-Barr wa al-Baḥr*. Leipzig: Otto Harrassowitz.

Arberry, A. J. 1983. *The Koran Interpreted*. London: Oxford University Press.

———. 1966. *Muslim Saints and Mystics. London*: Routledge and Kegan Paul.

———. 1963a. *Sufism*. London: Allen and Unwin.

Ardalan, Nader. 1980. "The Visual Language of Symbolic Form: A Preliminary Study of Mosque Architecture." In Jonathan G. Katz, ed., *Architecture as Symbol and Self-Identity*. Philadelphia: Aga Khan Award for Architecture, 18–36.

Ardalan, N., and L. Bakhtiar. 1973. *The Sense of Unity: The Sufi Tradition in Persian Architecture*. Chicago: University of Chicago Press, Phoenix ed., 1979.

ʿAṭṭār, Farīd al-Dīn. 1978. *The Conference of the Birds*, trans. C. S. Nott. London: Routledge and Kegan Paul.

Austin, R. W. J. 1980. *Ibn al-ʿArabī: The Bezels of Wisdom*. New York: Paulist.

———. 1971. *Sufis of Andalusia*. London: Allen and Unwin.

Al-Azraqī, Abū al-Walīd. 1983. *Akhbār Makka wa mā Jāʾa fīhā min al-Āthār*, ed. R. S. Malhas. Beirut: Dār al-Andalus.

Badawī, ʿAbd al-Raḥmān. 1978a. *Tārīkh al-Taṣawwuf al-Islāmī*. Kuwait: Wakālat al-Maṭbūʿāt.

———. 1978b. *Shakhṣiyyāt Qaliqa fī al-Islām*. Kuwait: Wakālat al-Maṭbūʿāt.

Bakhtiar, L. 1976. *Sufi: Expressions of the Mystic Quest*. London: Thames and Hudson, rep. 1979.

Al-Balāzurī. 1991. *Futūḥ al-Buldān*. Beirut: Dār al-Kutub al-ʿIlmiyya.

Behrens-Abouseif, Doris. 1999. *Beauty in Arabic Culture*. Princeton: Wiener.

Blair, Sheila, and Jonathan Bloom. 1994. *The Art and Architecture of Islam 1250–1800*. New Haven: Yale University Press.

Böwering, G. 1980. *The Mystical Vision of Existence in Classical Islam: The Hermeneutics of the Sufi Sahl at-Tusturī (d. 233/896)*. Berlin: De Gruyter.

Brookes, John. 1987. *Garden of Paradise: The History and Design of the Great Islamic Gardens*. London: Weidenfeld and Nicholson.

Burckhardt, Titus. 1986. *Alchemy*, trans. W. Stoddart. London: Element Books, 2nd imp. 1987.

———. 1983. *Universal Man*, trans. A. Culme-Seymour. Gloucestershire: Beshara.

———. 1979. "Concerning the '*Barzakh*'." *SCR* 13, 1–2: 24–30.

———. 1977. *Mystical Astrology according to Ibn ʿArabī*, trans. B. Rauf. Gloucestershire: Beshara.

———. 1976a. *An Introduction to Sufi Doctrine*, trans. D. M. Matheson. Northamptonshire: Thorsons.

———. 1976b. *The Art of Islam: Language and Meaning*. London: World of Islam Festival.

———. 1975. *The Wisdom of the Prophets (Fusus Al-Hikam)*, trans. A. Culme-Seymour. Gloucestershire: Beshara.

———. 1972. *Moorish Culture in Spain*. London: Allen and Unwin.

———. 1971. "Arab or Islamic Art? The Impact of the Arabic Language on the Visual Art." *SCR* 5, 1: 16–25.

———. 1970. "The Void in Islamic Art." *SCR* 4, 2: 96–99.

———. 1967a. *Sacred Art in East and West*. London: Perennial Books, 2nd imp. 1976.

———. 1967b. "Perennial Values in Islamic Art." *SCR* 1, 3: 132–41.

———. 1954. "The Spirit of Islamic Art." *IQ* 1, 1 (April): 210–18.

———. 1947. "Principles and Methods of Traditional Art." In B. Iyer, ed., *Art and Thought*. London: Luzac, 1947, 17–22.

Carpenter, E., and M. McLuhan. 1966. "Acoustic Space." In Carpenter and McLuhan, eds., *Exploration in* Communication. Boston: Beacon, 65–70.

Cassirer, Ernest. 1955. *The Philosophy of Symbolic Forms*. New Haven: Yale University Press.

Chittick, William C. 2000. *Sufism: A Short Introduction*. Oxford: Oneworld.

———. 1998. *The Self-Disclosure of God: Principles of Ibn al-ʿArabī's Cosmology*. Albany: State University of New York Press.

———. 1994. *Imaginal Worlds: Ibn al-ʿArabī and the Problem of Religious Diversity*. Albany: State University of New York Press.

———. 1989. *The Sufi Path of Knowledge: Ibn al-ʿArabī's Metaphysics of Imagination*. Albany: State University of New York Press.

Chodkiewicz, Michel. 1993. *An Ocean without Shore: Ibn ʿArabī, the Book, and the Law*. Albany: State University of New York Press.

Chorbachi, W. K. 1989. "In the Tower of Babel: Beyond Symmetry in Islamic Design." In I. Hargittai, ed., *Symmetry 2: Unifying Human Understanding*. Oxford: Pergamon, 751–89.

Cook, Roger. 1974. *The Tree of Life: Image for the Cosmos*. London: Thames and Hudson, rep. 1988.

Coomaraswamy, A. K. 1980a. "Symbols." *SCR* 14, 1–2: 15–17.

———. 1980b. "The Interpretation of Symbols." *SCR* 14, 1–2: 62–67.

———. 1979. *The Bugbear of Literacy*. Bedfont: Perennial Books.

———. 1977a. *On the Traditional Doctrine of Art*. Ipswich: Golgonooza.

———. 1977. *Selected Papers: 1 Traditional Art and Symbolism, 2 Metaphysics*, ed. Roger Lipsey. Princeton: Princeton University Press.

———. 1974. *Time and Eternity*. Ascona: Artibus Asiae.

———. 1972. *The Transformation of Nature in Art*. India: Monoharal.

———. 1970. "Khāwāj Khadir and the Fountain of Life in the Tradition of Persian and Mughal Art." *SCR* 4, 4: 221–30.

———. 1956. *Christian and Oriental Philosophy of Art*. New York: Dover.

Corbin, Henry. 1980a. "The Dramatic Element Common to the Gnostic Cosmogonies of the Religions of the Book." *SCR* 14, 3–4: 199–21.

———. 1980b. *Avicenna and the Visionary Recitals*, trans. W. R. Trask. Texas: Spring.

———. 1972. *Mudus Imaginalis or Imaginary and the Imaginal*. New York: Spring Books.

———. 1969. *Creative Imagination in the Sufism of Ibn Arabi*, trans. R. Manheim. Princeton: Princeton University Press.

———. 1965. "La Configuration du Temple de la Kaʿba comme Secret de la vie Spirituelle." *Eranos*, 79–166.

———. 1968. "The Force of Traditional Philosophy in Iran Today." *SCR* 2, 1: 12–26.

Craig, William L. 1979. *The Kalām Cosmological Argument*. London: Macmillan.

Crane, Howard, ed. 1987. *Risāle-i Miʿmāriyye: An Early-Seventeenth-Century Ottoman Treatise on Architecture*. Leiden: Brill.

Creswell, K. A. C. 1968. *Short Account of Early Muslim Architecture*. Beirut: Librairie du Liban.

Critchlow, K. 1976. *Islamic Patterns: An Analytical and Cosmological Approach*. London: Thames and Hudson, rep. 1983.

———. 1969. *Order in Space*. London: Thames and Hudson.

Daneshvari, A. 1986. *Medieval Tomb Towers of Iran*. Lexington: Mazda.

Danner, Victor. 1972. "The Necessity for the Rise of the Term Sūfī." *SCR* 6 (Spring): 71–77.

Davies, Paul. 1992. *The Mind of God: Science and the Search for Ultimate Meaning*. London: Simon and Schuster.

Dawood, N. J. 1983. *The Koran*. Middlesex: Penguin Books.

Daylamī, Abū al-Ḥasan. 1962. *Kitāb ʿAtf al-Alif al-maʾlūf ʿalā al-Lām al-Maʿtūf*, ed. J. C. Vadet. Cairo: Maṭbaʿat al-Maʿhad al-ʿIlmī al-Faransī li al-Āthār al-Sharqiyya.

Dickie, James. 1978. "Allah and Eternity: Mosques, Madrasas and Tombs." In Michell, ed., 1978, 15–47.

———. 1968. "The Hispo-Arab Garden: Its Philosophy and Function." *BSOAS* 31: 237–48.

Eliade, Mircea. 1994. *Al-Muqaddas wa al-ʿĀdī*, trans. ʿĀdel Al-ʿAwwā. Budapest: Sahara.

———. 1991. *Image and Symbols: Studies in Religious Symbolism*, trans. P. Mairet. Princeton: Princeton University Press.

———. 1986. *Symbolism, the Sacred, and the Arts*. New York: Crossroad.

———. 1974. *Patterns in Comparative Religion*, trans. R. Sheed. New York: New American Library.

———. 1971. *The Myth of the Eternal Return or, Cosmos and History*, trans. W. R. Trask. Bollingen: Princeton University Press, rep. 1974.

———. 1959a. "Methodological Remarks on the Study of Religious Symbolism." In Eliade and Kitagawa, eds., *The History of Religions: Essays in Methodology*. Chicago: University of Chicago Press, 7th imp. 1974, 86–107.

———. 1959b. *The Sacred and the Profane: The Nature of Religion*, trans. W. R. Trask. New York: Harcourt, Brace and World.

El-Said, I., and A. Parman. 1976. *Geometric Concepts in Islamic Art*. London: World of Islam Festival.

Esin, Emil. 1963. *Mecca the Blessed, Madinah the Radiant*. London: Elek Books.

Fakhri, Majid. 1983. *A History of Islamic Philosophy*. New York: Colombia University Press, 2nd ed.

Al-Fārābī, Abū Naṣr. 1968. *Kitab al-Jamᶜ bayn Raʾyyī al-Ḥakīmayn*. Beirut: Dār al-Mashriq.

Al-Farāhīdī, al-Khalīl b. Abī Aḥmad. 1405h. *Kitāb al-ʿAyn*, ed. M. Makhzūmī and A. al-Samirrāʾī. Qumm: Manshūrāt Dār al-Hijra.

Foster, Hal, ed. 1988. *Vision and Visuality*. Seattle: Dia Art Foundation.

Frishman, Martin, and Hasan-udin Khan, eds. 1994. *The Mosque: History, Architectural Development and Regional Identity*. London: Thames and Hudson.

Al-Ghazālī, Abī Ḥāmid. 1990. *Tahāfut al-Falāsifa*. Beirut: Dār al-Mashriq.

———. 1986a. *Al-Risāla al-Laduniyya*, ed. R. al-ᶜAbdullah. Damascus: Dār al-Ḥikma.

———. 1986b. *Majmūʿat Rasāʾil al-Imām al-Ghazālī*. Beirut: Dār al-Kutub al-ʿIlmiyya, 7 vols.

———. 1983a. *Iḥyāʾ ʿUlūm al-Dīn*. Beirut: Dār al-Maᶜrifa, 4 vols.

———. 1971. *Al-Maqṣad al-Asnā fī Sharḥ Maʿānī ʿAsmāʾ Allāh al-Ḥusnā*, ed. F. Shehadi. Beirut: Dār al-Mashriq.

———. nd. *Miʿyār al-ʿIlm*. Cairo: Dār al-Maᶜārif.

Gilsenan, M. 1973. *Saint and Sufi in Modern Egypt*. Oxford: Clarendon.

Gleiser, Marcelo. 2001. *The Prophet and the Astronomer: A Scientific Journey to the End of Time*. New York: Norton.

Grabar, Oleg. 1996. *The Shape of the Holy: Early Islamic Jerusalem*. Princeton: Princeton University Press.

———. 1992. *The Mediation of Ornament*. Princeton: Princeton University Press.

———. 1983. "Symbols and Signs in Islamic Architecture." In R. Holod and D. Rastorfer, eds., *Architecture and Community*. New York: Aperture, 25–32.

———. 1976. *Studies in Medieval Islamic Art*. London: Variorum Reprints.

———. 1973. *The Formation of Islamic Art*. New Haven: Yale University Press.

Grunebaum, G. E. Von. 1969. *Islam: Essays in the Nature and Growth of a Cultural Tradition*. London: Routledge and Kegan Paul.

Guénon, René. 1984. *The Multiple States of Being*, trans. J. Godwin. New York: Larson.

———. 1983a. *The Lord of the World*. England: Coombe Spring.

———. 1983b. *The Reign of Quantity and the Signs of the Times*, trans. Lord Northbourne. Pakistan: Suhel Academy Lahore.

———. 1980a. "The Corner-Stone." *SCR* 14, 3–4: 140–48.

———. 1980b. "The Arts and Their Traditional Conception." *SCR* 14, 3–4: 195–98.

———. 1975a. *The Crisis of the Modern World*, trans. M. Pallis and R. Nicholson. London: Luzac.

———. 1975b. *Symbolism of the Cross*, trans. A. Macnab. London: Luzac.

———. 1971a. "The Heart and the Cave." *SCR* 5, 1: 12–15.

———. 1971b. "The Mountain and the Cave." *SCR* 5, 2: 69–71.

———. 1969. "The Language of the Birds." *SCR* 3, 2: 94–97.

———. 1962. *Symboles Fondamentaux de la Science Sacrée*. Paris: Gallimard.

———. 1947. "The Mystries of the Letter Nūn." In Iyer, 1947, 166–68.

Al-Ḥallāj, Al-Ḥusain b. Manṣūr. 1994. *Al-Ṭawāsīn wa Bustān al-Maʿrifa*. Damascus: Dār al-Yanābīʿ.

———. 1997. *Al-Dīwān Yalīhi Kitāb al-Ṭawāsīn*. Beirut: Manshūrāt al-Jamal.

Hanaway, W. C., Jr. 1976. "Paradise on Earth: The Terrestrial Garden in Persian Literature." In MacDougall and Ettinghaausen, eds., 41–68.

Al-Ḥanbalī, Mujīr al-Dīn. 1999. *Al-Uns al-Jalīl bi-Tārīkh al-Quds wa al-Khalīl*, ed. A. Abū Tabbāna. Amman: Maktabat Dandīs, 2 vols.

Harley, J. B., and D. Woodward, eds. 1992. *History of Cartography II: Cartography in the Traditional Islamic and South Asian Societies*. Chicago: University of Chicago Press.

Harries, Karsten. 2001. *Infinity and Perspective*. Cambridge, Mass.: MIT Press.

Hartner, Willy. 1938. "The Pseudoplanetary Nodes of the Moon's Orbit in Hindu and Islamic Iconographies." *Ars Islamica* 5, 1: 113–54.

Heidegger, Martin. 1962. *Being and Time*, trans. J. Macquarrie and E. Robinson. London: Blackwell.

Hellinbrand, Robert. 1994. *Islamic Architecture: Form, Function and Meaning*. Edinburgh: Edinburgh University Press.

Hill, Derek, and O. Grabar. 1964. *Islamic Architecture and Its Decoration A.D. 800–1500*. London: Faber and Faber.

Hinen, Anton. 1982. *Islamic Cosmology*. Beirut: German Institute for Oriental Studies.

Hoag, John D. 1977. *Islamic Architecture*. New York: Abrams.

Hourani, George F., ed. 1975. *Essays on Islamic Philosophy and Science*. New York: State University of New York Press.

Al-Hujwīrī. 1976. *Kashf al-Maḥjūb*, trans. R. N. Nicholson. London: Luzac.

Ibn al-Athīr, ʿAlī b. Muḥammad. 1982. *Al-Kāmil fī al-Tārīkh*. Beirut: Dār Ṣadir, 13 vols.

Ibn ʿArabī, Muḥyī al-Dīn. 2002. *Rasāʾil Ibn ʿArabī*, ed. S. ʿAbd al-Fattāḥ. Beirut: al-Intishār al-ʿArabī, 3 vols.

———. 2000. *Majmūʿat Rasāʾil Ibn ʿArabi*. Beirut: Dār al-Maḥajja al-Bayḍāʾ, 2 vols.

———. 1985. *Shajarat al-Kawn*, ed. R. al-ʿAbdaullah. Beirut: Dār al-ʿIlm.

———. 1981. *Tafsīr al-Qurʾān al-Karīm,* ed. M. Ghālib. Beirut: Dār al-Andalus, 2 vols.

———. 1980. *Fuṣūṣ al-Ḥikam*, ed. A. Afīfī. Beirut: Dār al-Kitāb al-ʿArabī.

———. 1978. *The Tarjumān al-Ashwāq*, trans. R. A. Nicholson. London: Theosophical Publishing House.

———. 1976. *"Whoso Knoweth Himself . . . ,"* trans. T. H. Weir. Gloucestershire: Beshara.

———. 1971. The *Rūh al-Qudus* and *al-Durra al-Fākhira,* trans. R. W. Austin. London: Allen and Unwin.

———. 1970. *Rūḥ al-Qudus*. Damascus: ʿĀlam al-Fikr.

———. 1965. *Mawāqiʿ al-Nujūm wa Mutālaʿat Ahillat al-Asrār wa al-Nujūm*. Cairo: Muhḥammad ʿAlī Ṣabīḥ.

———. 1961. *Tanazzul al-Amlāk min ʿĀlam al-Arwāḥ ilā Ālam al-Aflāk*, ed. A. Srūr and A. Z. ʿAṭiyya. Damascus: Dār al-Fikr al-ʿArabī.

———. 1954. *ʿAnqāʾ Maghrib fī Khatm al-Awliyāʾ wa Shams al-Maghrib*. Cairo: Muḥammad ʿAlī Ṣabīḥ.

———. 1948. *Rasāʾil Ibn ʿArabī*. Hayderabad: Dāʾirat al-Maʾārif al-ʿUthmāniyya, 2 vols.

———. 1374h. *Tawajjuhāt al-Ḥurūf.* Cairo: Maṭbaʿat al-ʿIrfān.

———. 1336h. *Inshāʾ al-Dawāʾir*. Leiden: Brill.

———. 1336h. *ʿUqlat al-Mustawfiz*. Leiden: Brill.

———. 1336h. *Al-Tadbīrāt al-Ilāhiyya fī Iṣlāḥ al-Mamlaka al-Insāniyya*. Leiden: Brill.

———. nd. 1. *Al-Futūḥāt al-Makkiyya fī Maʿrifat al-Asrār al-Mālikiyya wa al-Malakiyya*. Beirut: Dār Ṣādir, 4 vols.

———. nd. 2. *Muḥāḍarat al-Abrār wa Musāmarat al-Akhyār fī al-Adabiyyāt wa al-Nawādir wa al-Akhbār*. Beirut: Dār Ṣādir, 2 vols.

Ibn al-ʿArīf, Abu al-ʿAbbās. 1980. *Maḥāsin al-Majālis*, trans. W. Elliott and A. K. Abdulla. Trowbridge: Avebury.

Ibn ʿAsākir, Abī al-Qāsim. 1995–2001. *Tārīkh Madīnat Dimashq*. Beirut: Dār al-Fikr.

Ibn al-Fāriḍ, ʿUmar. 1328h. *Dīwān Ibn al-Fāriḍ*. Beirut: Maṭbaʿit al-Tawfīq.

Ibn Kathīr, Abū al-Fidā' Ismā'īl.1985. *Al-Bidāya wa al-Nihāya*. Beirut: Maktabat al-Ma'ārif, 7 vols.

———. 1983. *Tafsīr al-Qur'ān al-'Aẓīm*. Beirut: Maktabat al-Ma'ārif, 4 vols.

Ibn Khaldūn, 'Abd al-Raḥmān. 1983. *Kitāb al-'Ibar wa Dīwān al-Mubtada' wa al-Khabar*. Beirut: Dār al-Kitāb al-Lubnānī, 14 vols.

Ibn Manẓūr, Muḥammad bin Mukarram. 1984. *Mukhtaṣar Tārīkh Dimashq li Ibn 'Asākir: Faḍā'il al-Shām wa Khiṭaṭ Dimashq*. Damascus, Dār al-Fikr.

Ibn al-Nadīm, Muḥammad b. Ishāq. 1978. *Al-Fihrist*. Beirut: Dār al-Ma'rifa.

Ibn Qanbar, Abī Bishr. nd. *Kitāb Sībawayh*, ed. A. M. Hārūn. Beirut: 'Ālam al-Kutub, 5 vols.

Ibn Qayyim al-Jawziyya. 1962. *Ḥādī al-Arwāḥ ilā Bilād al-Afrāḥ*. Cairo: Muḥammad 'Alī Ṣabīḥ, 4th ed.

Ibn Rushd, Abī al-Walīd Muḥammad. 1972. *Talkhīṣ Kitāb al-Ḥis wa al-Maḥsūs*, ed. H. Blumberg. Cambridge Mass.: Mediaeval Academy of America.

———. nd. *Kitāb Faṣl al-Maqāl wa Tafsīr mā bayn al-Sharī'a wa al-Ḥikma min al-Ittiṣāl*, ed. A. Abī Nādir. Beirut: Dār al-Mashriq.

Ibn Rusta, Aḥmad. 1891. *Kitāb al-A'lāq al-Nafīsa*. Leiden: Brill.

Ibn Sa'd, Muḥammad. nd. *Al-Ṭabaqāt al-Kubrā*. Beirut: Dār Ṣādir, 9 vols.

Ibn Sīnā, Abū 'Alī al-Ḥusayn. 1983. *Asbāb Ḥudūth al-Ḥurūf*, ed. M. H. al-Ṭayyān and Y. M. 'Alam. Damascus: Majma' al-Lugha al-'Arabiyya.

———. 1405h. *Al-Shifā'*. Qumm: Maktabat Āyat Allāh al-'Uẓmā al-Mar'ashī al-Najfī, 10 vols.

———. 1353h. *Rasā'il Ibn Sīnā*. Hayarabad: Dā'irat al-Ma'ārif al-'Uthmāniyya.

———. nd. *Al-Ishārāt wa al-Tanbīhāt*, ed. S. Dunyā. Cairo: Dār al-Ma'ārif, 3rd ed.

Ibn Taymiyya, Taqyyī al-Dīn. 2000. *Majmū' al-Fatāwā*, ed. A. Muṣtafā. Beirut: Dār al-Kutub al-'Ilmiyya, 21 vols.

Ibrāhīm, Maḥmūd. 1985. *Faḍā'il Bayt al-Maqdis fī Makhṭūṭāt 'Arabiyya*. Kuwait: Ma'had al-Makhṭūṭāt al-'Arabiyya.

Ikhwān al-Ṣafā'. 1984. *Al-Risāla al-Jāmi'a*, ed. M. Ghālib. Beirut: Dār al-Andalus.

———. nd. *Rasā'il Ikhwān al-Ṣafā' wa Khillān al-Wafā'*. Beirut: Dār Ṣādir, 4 vols.

Izutsu, Toshihiko. 1983. *Sufism and Taoism: A Comparative Study of Key Philosophical Concepts*. Tokyo: Iwanami Shoten.

———. 1971. *The Concept and Reality of Existence*. Tokyo: Keio Institute of Cultural and Linguistic Studies.

Jachimowicz, Edith. 1975. "Islamic Cosmology." In C. Blacker and M. Loewe, eds., *Ancient Cosmologies*. London: Allen and Unwin, 143–71.

Jamī, ʿAbd al-Raḥmān. 1914. *Lawāʾiḥ: A Treatise on Sufism*, trans. E. H. Whinfield and M. M. Kazvini. London: Royal Asiatic Society.

Jeffery, Arthur. 1959. "Ibn al-ʿArabī's Shajarat al-Kawn." *IS* 10: 43–77; 11: 133–60.

Al-Jīlānī, ʿAbd al-Qādir. 1353h. *Al-Maqṣad al-Asnā fī Naẓm Asmāʾ Allāh al-Ḥusnā*. Damascus: Maṭbaʿat al-Tawfīq.

———. 1281h. *Kitāb al-Fatḥ al-Rabbānī wa al-Fayḍ al-Rūḥānī*. Būlāq: Dār al-Ṭibāʿa.

Al-Jīlī, ʿAbd al-Karīm. 1340h. *Al-Kahf wa al-Raqīm fī Sharḥ Bism Allāh al-Raḥmān al-Raḥīm*. Hyderabad.

———. 1328h. *Al-Insān al-Kāmil fi Maʿrifat al-Awākhir wa al-Awāʾil*. Cairo: al-Maṭbaʿa al-Azhariyya.

———. nd. *Al-Isfār al-Gharīb Natījat al-Safar al-Qarīb*. Cairo: Dār al-Risāla.

———. nd. *Kitāb al-Nuqṭa*. Cairo: Dār al-Risāla.

Johnson, F. Ernest, ed. 1955. *Religious Symbolism*. New York: Institute for Religious and Social Studies.

Al-Jurjānī, ʿAlī. 1985. *Kitāb al-Taʿrīfāt*. Beirut: Dār al-Kitāb al-ʿArabī.

Karamustafa, Ahmet T. 1992. "Cosmographical Diagrams." In Harley and Woodward, eds., 1992, 71–89.

Al-Kāshānī, Afḍal al-Dīn. 1954. *Al-Muṣannafāt*, trans. S. H. Nasr. Tehran: Tehran University Press.

King, David A. 1999. *World-Maps for Finding the Direction and Distance to Mecca*. Leiden: Brill.

———. 1993. *Astronomy in the Service of Islam*. Hampshire: Variorum.

———. 1984. "The Astronomy of the Mamluks: A Brief Overview." *Muqarnas*, 2: 73–84.

King, David, and Richard Lorch. 1992. "*Qibla* Charts, *Qibla* Maps, and Related Instruments." In Harley and Woodward, eds., 1992, 189–205.

Knysh, Alexander. 2000. *Islamic Mysticism: A Short History*. Leiden: Brill.

———. 1999. *Ibn ʿArabī in the Later Islamic Tradition: The Making of a Polemical Image in Medieval Islam*. Albany: State University of New York Press.

Kolb, David. 1990. *Postmodern Sophistications: Philosophy, Architecture, and Tradition*. Chicago: University of Chicago Press.

Lancaster, Clay. 1956. "Metaphysical Beliefs and Architectural Principles." *JAAC* 14, 3: 287–303.

Lehrman, J. 1980. *Earthly Paradise*. Berkeley: University of California Press.

Lings, Martin. 1991. *Symbol and Archetype: A Study of the Meaning of Existence*. Cambridge: Quinta Essentia.

———. 1987. *The Quranic Art of Calligraphy and Illumination*. London, Scorpion, rep.

———. 1983. *Muḥammad.* London: Allen and Unwin.

———. 1979. "Sufi Answers to Questions on Ultimate Reality." *SCR* 13, 3–4: 137–47.

———. 1975. *What Is Sufism?* London: Allen and Unwin.

———. 1971. *A Sufi Saint of the Twentieth Century.* London: Allen and Unwin, 2nd ed.

———. 1968. "The Quranic Symbolism of Water." *SCR* 2, 3: 153–60.

———. 1965. *Ancient Beliefs and Modern Superstitions.* London: Perennial Books.

MacDougall, Elizabeth B., and Ettinghausen, Richard, eds. 1976. *The Islamic Garden.* Washington: Dumbarton Oaks.

McVey, Kathleen E. 1983. "The Domed Church as Microcosm: Literary Roots of an Architectural Symbol." *Dumbarton Oaks Papers* 37: 91–121.

Madkūr, Ibrāhīm, ed. 1969. *Al-Kitāb al-Tidhkārī li-Muḥyī al-Dīn Ibn ʿArabī.* Cairo: al-Maktaba al-ʿArabiyya.

Makārim, Sāmī. 1989. *Al-Ḥallāj: fī mā Warāʾ al-Maʿnā wa al-Khaṭṭ wa al-Lawn.* London: Riad el-Rayyes Books.

Al-Makkī, Abū Ṭālib. 2001. *Qūt al-Qulūb fī Muʿāmalāt al-Maḥbūb wa Waṣf Ṭrīq al-Murīd ilā Maqām al-Tawḥīd.* Cairo: Maktabat Dār al-Turāth.

———. 1964. *ʿIlm al-Qulūb.* Cairo: Maktabat al-Qāhira.

Al-Maqdisī, Shihāb al-Dīn. 1994. *Muthīr al-Gharām ilā Ziyārat al-Quds wa al-Shām.* Beirut: Dār al-Jīl.

Al-Maqrīzī, Aḥmad b. ʿAlī. 1996. *Kitāb al-Mawāʿiẓ wa al-Iʿtibār fī Dhikr al-Khiṭaṭ wa al-Āthār.* Cairo: Maktabat al-Ādāb, 4 vols.

Marmura, Michael E. 1984. *Islamic Theology and Philosophy: Studies in Honor of George F. Hourani.* Albany: State University of New York Press.

Massignon, L. 1982. *The Passion of al-Ḥallāj.* Princeton: Princeton University Press, 4 vols.

Massignon, L., and P. Kraus. 1936. *Akhbār al-Ḥallāj.* Paris: Larose.

Al-Masʿūdī, Abī ʿAbd al-Raḥmān. 1983. *Murūj al-Dhahab wa Maʿādin al-Jawhar*, ed. Y. A. Dāghir. Beirut: Dār al-Andalus, 4 vols.

Mayer, L. A. 1956. *Islamic Architects and Their Works.* Geneve: Kundig.

Meier, Fritz. 1976. "The Mystic Path." In B. Lewis, ed., *The World of Islam: Faith People and Culture.* London: Thames and Hudson, 117–39.

Michell, George, ed. 1978. *Architecture of the Islamic World.* London: Thames and Hudson.

Michon, L. J. 1982. "Education in the Traditional Arts and Crafts and the Cultural Heritage of Islam." In Nasr, ed., 1982, 49–62.

Al-Muḥāsibī, Abū ʿAbd Allāh. 1940. *Kitāb Al-Riʿāya li-Ḥuqūq Allāh*, ed. M. Smith. London: Luzac.

Al-Muqaddasī, Abū ʿAbd Allāh. 1987. *Aḥsan al-Taqāsīm fī Maʿrifat al-Aqālīm*. Beirut: Dār Iḥyāʾ al-Turāth al-ʿArabī.

Murata, Sachiko. 1987. "The Angels." In Nasr, ed., 1987d, 324–44.

Al-Nābulusī, ʿAbd al-Ghanī. 1995. *Al-Wujūd al-Ḥaqq*, ed. B. Aladdin. Damascus: Institut Francais de Damas.

———. nd. *Al-Qawl al-Matīn fī Bayān Tawḥīd al-ʿĀrifīn*. Cairo: Muḥammad ʿAlī Ṣabīḥ.

Nasr, Seyyed Hossein. 1987a. *Islamic Art and Spirituality*. Ipswich: Golgonooza.

———. 1987b. *Traditional Islam in the Modern World*. London: KPI .

———. 1981. *Knowledge and the Sacred*. New York: Crossroad.

———. 1980. *Living Sufism*. London: Unwin Paperbacks.

———. 1979. *Ideals and Realities of Islam*. London: Unwin Paperbacks.

———. 1978a. *An Introduction to Islamic Cosmological Doctrines*. London: Thames and Hudson, rev. ed.

———. 1978b. *Ṣadr al-Dīn Shīrāzī and His Transcendent Theosophy*. Tehran: Imperial Iranian Academy of Philosophy.

———. 1976a. *Man and Nature: The Spiritual Crisis of Modern Man*. London: Unwin Paperbacks, Mandala ed.

———. 1976b. *Islamic Science: An Illustrated Study*. London: World of Islam Festival Trust.

———. 1975. *Islam and the Plight of Modern Man*. London: Longman.

———. 1972. "Mullā Ṣadrā and the Doctrine of Unity of Being." *Philosophical Forum* 4, 1: 153–61.

———. 1968. *Science and Civilization in Islam*. Cambridge, Mass.: Harvard University Press.

———. 1964. *Three Muslim Sages*. Cambridge, Mass.: Harvard University Press.

Nasr, Seyyed Hossein, ed. 1987d. *Islamic Spirituality: Foundations*. New York: Crossroad.

———. 1982. *Philosophy, Literature and Fine Arts*. Jeddah: King Abdoulaziz University.

Necipoğlu, Gülru. 1995. *The Topkapi Scroll: Geometry and Ornament in Islamic Architecture*. Santa Monica: Getty Center for the History of Art and the Humanities.

Netton, Ian R. 1982. *Muslim Neoplatonists*. London: Allen and Unwin.

Nicholson, R. A. 1966. *The Mystics of Islam*. London: Routledge and Kegan Paul.

———. 1911. *Al-Hujwīrī: Kashf al-Muḥjūb*. London: Luzac.

Al-Niffarī, Muḥammad b. ʿAbd al-Jabbār. nd. *Kitāb al-Mawāqif*, ed. M. ʿAbbūd. Beirut: Dār al-ʿĀlam al-Jadīd.

Ong, Walter J. 1977. *Interfaces of the Word*. Ithaca: Cornell University Press.

———. 1967a. *In the Human Grain*. New York: Macmillan.

———. 1967b. *The Presence of the Word*. New Haven: Yale University Press.

l'Orange, H. P. 1973. *Likeness and Icon: Selected Essays in Classical and Early Mediaeval Art*. Odense: Odense University Press.

Otto, Rudolph. 1957. *Mysticism East and West*. New York: Meridian Books.

———. 1926. *The Idea of the Holy*, trans. J. Harvey. London: Oxford University Press, 4th ed.

Paccard, André. 1980. *Traditional Islamic Craft in Moroccan Architecture*, trans. M. Guggenheim. Saint-Jorioz: Editions Ateliers 74, 2 vols.

Palacios, M. Asin. 1979. *Ibn ʿArabī: Ḥayātuhu wa Madhhabuhu,* trans. ʿAbd al-Raḥmān Badawī. Kuwait: Wakālat al-Maṭbūʿāt.

Papadopoulo, Alexander. 1980. *Islam and Muslim Art*. London: Thames and Hudson.

Péres-Gómez, Alberto, and Louis Pelletier. 1997. *Architectural Representation and the Perspective Hinge*. Cambridge: MIT Press.

Pickthall, Marmaduke. 1930. *The Meaning of the Glorious Koran*. London: Knopf.

Al-Qāshānī, ʿAbd al-Razzāq. nd. *Shraḥ al-Qāshānī ʿalā Fuṣūṣ al-Ḥikam*. Egypt: al-Maṭbaʿa al-Maymaniyya.

Al-Qūnawī, Ṣadr al-Dīn. 1983. *Mirʾāt al- ʿĀrifīn,* trans. S. H. Askari as *Reflection of the Awakened*. London: Zahara Trust.

Al-Qushayrī, Abī al-Qāsim. nd. *Al-Risāla al-Qushayriyya fī ʿIlm al-Taṣawwuf.* Beirut: Dār al-Kitāb al-ʿArabī.

Rahman, F. 1964. "Dream, Imagination and ʿĀlam al-Mithāl." *IS* 3: 167–80.

Raphael, Melissa. 1997. *Rudolf Otto and the Concept of Holiness*. Oxford: Clarendon.

Al-Rāzī, Fakhr al-Dīn. 1990. *Al-Mabāḥith al Mashriqiyya fī ʿIlm al-Ilāhiyyāt wa al-Ṭabīʿiyyāt*. Beirut: Dār al-Kitāb al-ʿArabī.

———. 1986. *Al-Nafs wa al-Rūḥ*. Damascus: Dār al-Ḥikma.

Reat, Noble Ross. 1975. "The Tree Symbol in Islam." *SCR* 9, 3: 164–82.

Rees, Martin. 2001. *Our Cosmic Habitat*. Princeton: Princeton University Press.

Ritter, Hellmut. 2003. *The Ocean of the Soul: Man, the World, and God in the Stories of Farīd al-Dīn al-ʿAṭṭār*, trans. J. O'Kane. Leiden: Brill.

Rizq, ʿĀsim M. 1997. *Khanqāwāt al-Ṣūfiyya fī Miṣr fī al-ʿAṣrayn al-Ayyūbī wa al-Mamlūkī (567–923h /1171–1517)*. Cairo: Maktabat Madbūlī, 2 vols.

Rosenthal, F. 1971. *Four Essays on Art and Literature in Islam*. Leiden: Brill.

Rūmī, Jalāl al-Dīn. 1925. *The Mathnawī of Jalālu-ddīn Rūmī*, trans. R. Nicholson. London: Luzac, 3 vols.

Safadi, Yasin Hamid. 1978. *Islamic Calligraphy*. London: Thames and Hudson.

Saliba, A. 1976. *Homo Religious in Mircea Eliade: An Anthropological Evaluation*. Leiden: Brill.

Al-Sarrāj, Abū Naṣr ʿAbd Allāh. 1963. *Kitāb al-Lumaʿ fī al-Taṣawwuf*, ed. R. A. Nicholson. London: Luzac.

Schimmel, Annemarie. 1994. *Deciphering the Signs of God: A Phenomenological Approach to Islam*. Edinburgh: Edinburgh University Press.

———. 1984. *Calligraphy and Islamic Culture*. Albany: New York University Press.

———. 1978. *Mystical Dimensions of Islam*. Chapel Hill: University of North Carolina Press, 3rd ed.

———. 1976. "The Celestial Garden in Islam." In MacDougall and Ettinghusen, 1976, 13–16.

Schuon, Frithjof. 1986. *Survey of Metaphysics and Esoterism*, trans. G. Polit. Bloomington: World Wisdom Books.

———. 1985. *Christianity/Islam: Essays on Esoteric Ecumenicism*, trans. G. Polit. Bloomington: World Wisdom Books.

———. 1984a. *Light on the Ancient Worlds*, trans. L. Northbourne. Bloomington: World Wisdom Books, 2nd ed.

———. 1984b. *The Transcendent Unity of Religions*, trans. H. Smith. Wheaton: Theosophical.

———. 1982a. *Castes and Races*, trans. M. Pallis and M. Matheson. Bedfont: Perennial Books.

———. 1982b. *From the Divine to the Human*, trans. G. Polit and D. Lambert. Bloomington: World Wisdom Books.

———. 1981a. *Esoterism as Principle and as Way*, trans. W. Stoddart. Bedfont: Perennial Books.

———. 1981b. *Sufism: Veil and Quintessence*, trans. W. Stoddart. Bloomington: World Wisdom Books.

———. 1980. *Stations of Wisdom*, trans. G. E. Palmer. Bedfont: Perennial Books.

———. 1979b. "Sophia Perennis." *SCR* 13, 3–4: 133–37.

———. 1976a. *Islam and the Perennial Philosophy*, trans. J. P. Hobson. London: World of Islam Festival.

———. 1976b. *Understanding Islam*, trans. D. M. Matheson. London: Unwin Paperbacks, rep. 1981.

———. 1975. *Logic and Transcendence*, trans. P. N. Townsend. Bedfont: Perennial Books, rep. 1984.

———. 1969. *Dimensions of Islam*, trans. P. N. Townsend. London: Allen and Unwin.

———. 1959b. *Language of the Self*. Madras: Baneshan.

———. 1959c. *Gnosis, Divine Wisdom*. London: Murray.

———. 1954. *Spiritual Perspectives and Human Facts*. London: Faber and Faber.

Shehadi, Fadlou A. 1975. "Arabic and the Concept of Being." In G. Hourani, ed., *Essays on Islamic Philosophy and Science*. Albany: State University of New York Press, 147–57.

Sherrard, Philip. 1990. *The Sacred in Life and Art*. Ipswich: Golgonooza.

———. 1980. "Art and Originality." *SCR* 14, 3–4: 167–77.

Shīrāzī, Ṣadr al-Dīn. 2002. *Al-Ḥikma al-Muta'āliya fī al-Asfār al-'Aqliyya al-Arba'a*. Beirut: Dār Ihyā' al-Turāth al-'Arabī, 8 vols.

Smith, Earl Baldwin. 1978. *Architectural Symbolism of Imperial Rome and the Middle Ages*. New York: Hacker Art Books.

———. 1950 *The Dome: A Study in the History of Ideas*. Princeton: Princeton University Press.

Smith, Margaret. 1954. *The Sufi Path of Love*. London: Luzac.

———. 1944. *Al-Ghazali, the Mystic*. London: Luzac.

———. 1938. "Al-Risālah al-Laduniyyah." *JRAS* (April), 2: 177–200; (July), 3: 353–74.

Snodgrass, Adrian. 1990. *Architecture, Time and Eternity: Studies in the Stellar and Temporal Symbolism of Traditional Buildings*. New Delhi: International Academy of Indian Culture.

———. 1985. *The Symbolism of the Stupa*. Ithaca: Cornell University Press.

Sönmez, Zeki. 1988. *Mimar Sinan Ile Ilgili Tarihi* Yazmalar-Belgeler. Istanbul: Mimar Sinan University.

Stewart, Desmond. nd. *Mecca*. New York: Newsweek.

Stross, Brian. 1994. "The Sacrum Bone: Doorway to the Otherworld." www.utexas.edu/courses/stross/papers/sacrum.

Sultan, A. A. 1980. "Notes on the Divine Proportions in Islamic Architecture." *Process Architecture* 15 (May): 131–63.

Al-Suyūṭī, Abī 'Abd Allāh. 1983. *Itḥāf al-Akhiṣṣā bi-Faḍā'il al-Masjid al-Aqṣā,* ed. A. R. Aḥmad. Cairo: al-Hay'a al-Miṣriyya al-'Āmma li al-Kitāb, 2 vols.

Al-Suyūṭī, Jalāl al-Dīn 'Abd al-Raḥmān. 1982. *Al-Hay'a al-Saniyya fī al-Hay'a al-Suniyya*, ed. A. Hinen. Beirut: German Institute for Oriental Studies.

Al-Ṭabarī, Abī Ja'far. 1992. *Jāmi' al-Bayān fī Ta'wīl al-Qur'ān*. Beirut: Dār al-Kutub al-'Ilmiyya.

———. nd. *Tārīkh al-Umam wa al-Mulūk*. Beirut: Maktabat Khayyāṭ, 6 vols.

Tabba, Yasser. 2001. *The Transformation of Islamic Art and Architecture during the Sunni Revival*. Seattle: University of Washington Press.

———. 1988. "Geometry and Memory in the Design of the Madrasat al-Firdaws in Aleppo." In *Theories and Principles of Design in the Architecture of Islamic Societies*. Cambridge: Aga Khan Program for Islamic Architecture, 23–34.

Tibawi, A. L. 1956a. "Ikhwān aṣ-Ṣafā' and Their Rasā'il: A Critical Review of a Century and a Half of Research." *IQ* 2: 28–46.

Tillich, Paul. 1961a. "The Meaning and Justification of Religious Symbols." In Hook, 1961, 3–11.

———. 1961b. "The Religious Symbol." In Hook, 1961, 301–21.

———. 1959. *Theology and Culture*. Oxford: Oxford University Press.

———. 1955. "Theology and Symbolism." In Johnson, 1955, 107–16.

Al-Tirmidhī, Muḥammad b. ʿAlī. 1975. *Kitāb Khatm al-Awliyā'*, ed. U. I. Yaḥyā. Beirut: Imprimerie Catholique.

Trimingham, J. Spencer. 1971. *The Sufi Orders in Islam*. Oxford: Clarendon.

Tuft, Anthony K. 1981. "Symbolism and Speculation in Suhrawardi's *The Song of Gabriel's Wing*." In Morewedge, 1981, 207–21.

Tusturī, Sahl b. ʿAbd Allāh. 1908. *Tafsīr al-Qur'ān al-ʿAẓīm*. Misr: Maṭbaʿat al-Saʿāda.

Vitruvius. 1960. *The Ten Books on Architecture*, trans. M. H. Morgan. New York: Dover.

Al-Walī, Ṭāhā. 1988. *Al-Masājid fī al-Islām*. Beirut: Dār al-ʿIlm li al-Malāyīn.

Al-Wāsiṭī, Abū Bakr. 1979. *Faḍā'il al-Bayt al-Muqaddas*, ed. I. Hassoun. Jerusalem: Magnes.

Wasserstrom, Steven. 1999. *Religion after Religion*. Princeton: Princeton University Press.

Wendell, C. 1971. "Baghdad: Imago Mundi." *IJMES* 2: 99–116.

Wensinck, A. J. 1978. *Studies of A. J. Wensinck*. New York: Arno.

Wensinck, A. J. (Junior). "Kaʿba." *EI2*.

Wertheim, Margaret. 1999. *The Pearly Gates of Cyberspace: A History of Space from Dante to the Internet*. Sydney: Doubleday.

Whitehead, A. N. 1928. *Symbolism: Its Meaning and Effect*. Cambridge: Cambridge University Press.

Wright, R. Ramsey. 1934. *Al-Bīrūnī: Elements of Astrology*. London: Luzac.

Wu, Kuan-Ming. 1998. *On the Logic of Together-ness: A Cultural Hermeneutic*. Leiden: Brill.

Yaḥyā, ʿUthmān. 2001. *Mu'allafāt Ibn ʿArabī: Tārīkhuhā wa Taṣnīfuhā*. Cairo: al-Hay'a al-Miṣriyya al-ʿĀmma li al-Kitāb.

Al-Yaʿqūbī, Aḥmad. 1891. *Kitāb al-Buldān*. Leiden: Brill.

———. nd. *Tārīkh al-Yaʿqūbī*. Beirut: Dār Ṣādir.

Al-Zamakhsharī, Abī al-Qāsim. 1979. *Asās al-Balāgha*. Beirut: Dār al-Maʿrifa.

Al-Ziriklī, Khayr al-Dīn. 1992. *Al-Aʿlām*. Beirut: Dār al-ʿIlm li al-Malāyīn.

Index